SONOGRAPHS OF THE SEA FLOOR

A Picture Atlas

SONOGRAPHS OF THE SEA FLOOR

A Picture Atlas

Frontispiece: Preparing to submerge the towed long-range sonar vehicle (GLORIA). This houses the transducer array, while the amplifier and the control and recording unit are on board ship, and connected via the towing cable.

SONOGRAPHS OF THE SEA FLOOR

A Picture Atlas

R.H. BELDERSON
N.H. KENYON
A.H. STRIDE
A.R. STUBBS

National Institute of Oceanography,
Wormley, Godalming , Great Britain

ELSEVIER PUBLISHING COMPANY Amsterdam, London, New York 1972

ELSEVIER PUBLISHING COMPANY
335 JAN VAN GALENSTRAAT
P.O. BOX 211, AMSTERDAM, THE NETHERLANDS

AMERICAN ELSEVIER PUBLISHING COMPANY, INC.
52 VANDERBILT AVENUE
NEW YORK, NEW YORK 10017

Library of Congress Card Number: 79-179997

ISBN 0-444-40984-X

With 163 illustrations

Printed in The Netherlands

PREFACE

Side-scan sonar has been used in the sea by research establishments and by industry during the past 15 years, but the relatively few interpretations of the acoustic patterns that have been published are scattered in a variety of scientific articles. The present book provides a summary of patterns so far recognised on side-scan sonar records and offers interpretations of them in geological and other terms. Most of the sonographs are drawn from the extensive collection at the British National Institute of Oceanography, but some have been supplied from elsewhere. Although the book is written primarily for the geologist it is thought it will be of value to other workers concerned with the sea floor and will appeal to teachers and to the general public for whom the sea floor holds the fascination of an unknown 'world'. For example, the geologist will find sonographs of many of the rock structures he sees on land, as well as a variety of sedimentary patterns and elements of relief which are also common in rivers and in sandy deserts. The hydraulics engineer can see illustrations of new or known bed forms characteristic of sediment transport, which have reached a size and development that is not possible in a flume or river because of their relatively limited dimensions. For the civil engineer there are examples of mobile or stable floors, and smooth or rough ones of relevance to the choice of routes for submarine cables and pipelines and for the sites of drilling rigs and production platforms; the method provides a ready means of searching for cables, pipelines and navigational hazards. In addition, there are sonographs of value to ecologists and to the somewhat conflicting dredging and fishery interests. Simplified explanations of some of the less common technical terms used in this book are given in a short glossary.

A variety of names have been given to the side-scan sonars now in use. The earliest was 'Asdic', due to the similarity between it and the submarine detection equipments used during World War II. Other names include: basdic, sideways asdic, sideways looking sonar, sideways sonar, echo-ranger, horizontal echo-sounder and lateral echo-sounder. Most of these are fairly clumsy and the last two are a conflict of terms, as echo-sounders measure the *depth* of the sea. The authors prefer the name of *side-scan sonar*, and describe the records as *sonographs*.

The authors wish to thank Mr. N.T. Mansbridge and Mr. M.J. Conquer for their able work in preparing the photographs; Mrs. C.E. Darter for her care in preparing the line drawings; Mr. M.J. Tucker and Dr. J.B. Wilson for suggesting improvements to the text. They are grateful to numerous colleagues for their hard work both at sea and in the design and making of side-scan sonar equipment. The generosity of other workers has made available some excellent sonographs of special topics and provided valuable records from far afield: Fig. 146, 148 and 149 were supplied by the Fisheries Laboratory, Lowestoft; Fig. 18 by Institut Français du Pétrole; Fig. 19, 28, 81, 82, 84, 129 and 141 by the Geophysics Group of Bath University and the Physics Dept. of Hong Kong University; Fig. 85, 157 and 158 by British Petroleum Co. Ltd.; Fig. 155, 156 and 159 by Decca Survey Ltd.; Fig. 79, 80, 160 and 161 by Hunting Surveys and Consultants Ltd., and Fig. 144, 145, 153, 154 and 162 by Kelvin Hughes (Smith Industries) Ltd. It is also a pleasure to thank Mr. K. R. Honick of the Royal Aircraft Establishment at Farnborough for going to so much trouble to convert some of our records to an almost true plan view, by means of his continuous flow copying machine.

R.H. Belderson
N.H. Kenyon
A.H. Stride
A.R. Stubbs
(1971)

CONTENTS

INTRODUCTION

The side-scan sonar method

Light and radar waves, so valuable for mapping the surface of the land, are too easily absorbed under water to be of more than local use; however, sound provides a valuable alternative. The side-scan sonar method can be considered as analogous to aerial photography when the sun is low and behind the camera, although there are marked differences in geometry, resolution and information content. This method produces a plan view of the shape and texture of the surface of the sea floor out to a maximum range at the present time of 22 km from the ship, and provides a basis for tying together detailed information obtained by sampler and camera at isolated points and by echo-sounder and sub-bottom profiler along widely spaced lines. The physical background of sonar is described by D.G. Tucker (1966) and a more detailed account is given by Urick (1967). More recently the side-scan principle has been used effectively with airborne radar to determine the geology of the land, e.g., Reeves (1969).

The principle of the side-scan sonar method is quite simple (see schematic diagram, page 6). Short pulses of sound are transmitted and echoes received by a transducer pointing sideways from a ship. The pulses are transmitted at regular intervals of time and the resulting echoes, from ridges of rock for instance, are usually displayed on a paper recorder such that echoes from near ridges are recorded first and echoes from more distant ridges are recorded progressively later. The groups of echoes from each pulse are displayed by a stylus moving across the paper so as to build up a picture line by line as the vessel advances. By placing these lines close together successive echoes from a ridge will appear to be continuous and indicate its form and location with respect to the ship.

The clarity or resolution of the sonograph depends on the narrowness of the sound beam in the direction of the ship's track and on the duration of the outgoing pulse (the pulse length) in the direction at right angles to the track. Typical values are $2°$ for the horizontal beam angle and 1 msec (1.5 m) for the pulse length in an instrument with a range of 1,000 m. The beam angle in the vertical plane is less important. Some instruments use wide beams which can be produced by small unstabilised transducers, but narrower beams, typically $10°$ wide, can be of value to reach the more distant ranges by concentrating the power, but more important, the ground between the near edge of this main beam and the floor beneath the ship is explored by a number of side lobes (weaker secondary beams) which reach the floor at angles of up to $90°$. Because their range is short and the angle of incidence steep, these side lobes produce quite strong echoes (see page 15). Side lobes can be of considerable value as they show the existence and trend of relief in the oblique profiles parallel to the ship's track. They also help in distinguishing patterns produced by midwater scatterers (for example fish shoals) from patterns on the sea bed (see page 3). However, information is inevitably lost due to the dead spaces between them, and for maximum pictorial effect it is better to have no side lobes at all. The edge of the sonograph beneath the ship (the profile) may then be visualised as analogous to a horizon.

Most recorders employ a fixed range or ranges so that for a given depth of water there is an optimum angle of depression of the main sound beam for 'illuminating' the sea floor and making the most effective use of the recorder width. This suggests that the tilt angle should be continuously varied with changes in water depth, although in practice the sonographs are easier to interpret if the changes are kept to a minimum, and if for irregular floors the beam is depressed in accordance with the mean depth. Other effects resulting from an increase in water depth are: that for a given transducer depth there is a corresponding increase in the width of blank paper between the transmission and the profile P; shadows shorten and the width scale of the sonograph becomes nearer that for a true plan view because of the steeper angle at which the sound reaches the floor, and, finally, as the angle becomes quite steep, the resolution in range becomes poorer because it is becoming increasingly dependent on the vertical angle of the beam. If really high resolution in range is required, so as to detect boulders, for instance, then a high frequency sound must be used, although this will severely limit the maximum range obtained. Typically, a resolution in range of 15 cm is accompanied by a maximum range of only about 300 m. On the other hand, a long range of say 22 km is associated with a resolution in range of no better than 7 metres. In practice, various factors limit the resolution to about one thousandth of the range.

Echoes from objects close to the ship are much stronger than those originating from further away. Because of this, and because most recorders cannot handle the full variation in the strength of the echoes, the amplification in the receiver increases with the increasing distance of their point of origin from the ship (Time Varied Gain). It is common practice to use a paper recorder to give an immediate-appreciation sonograph, while an oscilloscope or amplitude recorder will give a quantitative measure of the strength of the returning echoes which can be a means of identifying sediment grain size in ideal circumstances. There are also advantages in recording on magnetic tape as well, for signal processing during replay can sometimes produce sonographs of higher quality than those obtained in the first instance.

Development of the equipment

The idea of employing the acoustic side-scan principle for geological purposes was first described in the late 1950s. Kunze (1957) towed a pair of transducers so as to examine the sea floor on both sides of a ship at the same time. Chesterman et al. (1958) used a single ship-mounted transducer which was stabilised for the roll and yaw of the ship and in which the angle of depression of the beam could be altered remotely. Subsequent ship-borne equipment, made by the British National Institute of Oceanography, was described by Tucker and Stubbs (1961). Kelvin Hughes (1960) produced the 'Fishermans Asdic' which could be used as a side-scan sonar and they also made a commercial version of the Institute's equipment (Kelvin Hughes, 1962). Later developments of the more conventional side-scan sonar include the presentation of the sonograph with equal length and width scales (Cholet et al., 1968; Hopkins, 1970) and also automatic correction for the crabwise motion of a ship through the water induced by adverse sea or wind (Hopkins, 1970). More sophisticated devices, providing similar information, are the electronic sector-scanning sonars (Tucker et al., 1959; Voglis and Cook, 1966). The A.R.L. electronic sector-scanner uses a beam only 1/3° wide in the horizontal direction which is scanned rapidly over an arc of 30°. The sector-scanner has the advantage over fixed-beam sonars that it can be used to see features in any direction with respect to the ship, even when the ship is stationary, and can also be used to watch features in motion such as fish shoals. At the present stage this information is presented on an oscilloscope screen and recorded on cine-film.

The side-scan principle has been adapted for use in the deep sea in two main ways. The first method has been to make use of short range equipment by towing a pair of transducers relatively close to the floor on the end of a long cable (for instance, Clay and Liang, 1964; Laing and Nelkin, 1966). Such equipment has a relatively short range, towing speed is only a knot or so and there is a chance that it may get damaged or even lost through collision with the floor. However, such an approach provides considerable detail about the floor and the towed vehicle has also served as a satisfactory housing for other geophysical equipment, as described in the review by Spiess and Mudie (1970). Side-scan sonar has also been carried aboard the bathyscaphe "Archimede" (Brakl et al., 1969). The second approach (Rusby, 1970) has been to develop a much longer range equipment, called **GLORIA** (from Geological LOng Range Inclined Asdic), so as to operate from near the sea surface. This necessitates a low frequency, in order to obtain a long range, and this in turn calls for a large transducer (see frontispiece) because of both the power and the resolution in horizontal angle that are required. The worst of the sound refraction effect, which otherwise would limit the range, is overcome by towing the transducer a few hundred feet down so as to place it below the steepest part of the temperature gradient (thermocline). The resolution is inevitably less than for short range equipment, but the range of up to 22 km and the towing speed of up to 7 knots are obvious advantages.

In practice, no single device will meet the conflicting requirements of range and resolution, so that a variety of side-scan sonars are required if the whole of the sea and ocean floors are to be explored by this method.

Some applications

The first published account of the value of sonographs for indicating the form and composition of sea floors appeared in 1958 (Chesterman et al., 1958) although faulted rock outcrops had been recognised in the early 1950s. This was followed by use of the equipment to make a detailed geological map of a rock floor (Donovan and Stride, 1961a), and then to determine sediment transport paths on the floor (Stride, 1963), the zonal distribution of mobile bed forms on these paths (Belderson and Stride, 1966) and to recognise variation in the shape of bed forms, such as sand waves (Kenyon and Stride, 1968), sand ribbons (Kenyon, 1970a), longitudinal furrows (Dyer, 1970), and submarine canyons (Belderson and Stride, 1969), to give a few examples. At the present time publications show that the equipment has been used extensively only in European waters, although it has proved of value on similar floors near Hong Kong and has been used on such additional features as coral

(Wong et al., 1970) and large boulders (Chesterman et al., 1967). American workers were the first to use the equipment to examine the deep sea floor, stimulated as they were by the loss of the submarine "Thresher" (Spiess and Maxwell, 1964). They recognised a pattern of sedimentary lineations (Clay et al., 1964), and possible volcanic extrusions and intrusions (Spiess et al., 1969), studied the valleys on the depositional fans below submarine canyons (Mudie et al., 1970) and located small volcanic knobs on abyssal hills (Luyendyk, 1970). A review of many of these was given by Spiess and Mudie (1970). Long range side-scan (GLORIA) has been used to examine volcanoes and seamounts, as well as channels, canyons and slumps (Belderson et al., 1970; Stride, 1970)

Non-geological uses of this equipment have included the search for navigational hazards such as wrecks, the sides of channels (Kunze, 1957) and peaks of rock (Chesterman et al., 1967), as well as the search for fish shoals (Harden Jones and McCartney, 1962), the study of spawning grounds (Stubbs and Lawrie, 1962), the search for a Roman wreck (McGehee et al., 1968) and as a navigation technique (Lowenstein, 1970). A review of some of these applications, and of certain other effects, was given by Stubbs (1963). Electronic sector-scanning sonar has provided details about wrecks, areas dredged for gravel and about sediment bed forms, but is of most value for watching the movement of features in the water or on the floor, such as fish and fish shoals, and fishing gear in operation.

Interpretation of sonographs

Interpretation is taken as far as possible with present knowledge, although the authors realise the need for considerably more information about certain patterns. However, to limit the patterns illustrated to those which could be interpreted with certainty would entail leaving out a number of interesting ones which further research may prove to be of particular significance. More generally, it must be appreciated that there is some uncertainty in the descriptions of many of the patterns shown, for each sonograph refers to a large area of floor whose complexity frequently defies full examination by any other available method.

The numerous patterns displayed on the sonographs were differentiated from one another in the first instance by searching through the extensive collection of records for an example of each type whose essential characteristics were best developed and occurred on their own. The successful interpretation of these pat-terns usually necessitated a knowledge of the characteristics of the equipment used, as well as the sea conditions and a broad knowledge of the locality. For example, rock outcrop patterns on the sea floor have been followed up to the beach, dips and strikes have been determined by divers or else by echo-sounding and sub-bottom profiling gear, and numerous rock samples have been taken. Sediments have been collected by sampling from surface ships, submersible vehicles and by divers. Observations have been made by underwater television and also by cameras which can be towed behind a ship. Inferences about the nature of the material seen on the sonographs have been drawn by analogy with the results of flume experiments on the same type of bed forms. Use has been made of the abundant notations describing the nature of the sea floor that are given on navigational charts. Fish and plankton are indicated as midwater sound scatterers by means of echo-sounders as well as being recorded between bottom echoes produced by the side lobes of side-scan sonars, while visual observations of marine mammals have been correlated with effects shown on the sonograph. Similarly, with experience at sea the echoes from surface waves, ships and their wakes, as well as propeller noise, can usually be identified. The Lloyd Mirror effect is well known and easily recognised. Sound refraction effects are known to be common in stratified seas and are complicated by internal waves: they provide an acceptable explanation for some striking patterns on sonographs for which there is no other known origin.

The most abrupt relief of the sea floor is brought out by the shadows it casts. The relative length of the shadow increases with distance from the transducer because of the increasingly oblique approach of the sound. Positive and negative elements of relief may be distinguished by the relative location of the shadow and high light with respect to the sound source. Features with asymmetric profiles, such as sand waves and rock outcrops, will have a different (sometimes greatly different) appearance according to the direction from which they are viewed. Slopes facing the ship will give a strong reflection and slopes facing away from the ship, if not in shadow, will give a poor reflection.

The texture of the floor also affects the strength of the echoes. Only the small portion of the incident sound which is scattered back to the ship will be recorded, the remainder going off in other directions. Mud and sand will tend to cause little backscattering towards the ship but shells, gravel and coarser grades will send back much more of the incident sound, each large grain making a contribution to a generally higher 'reverberation' level. However, only large individual components such as boulders can be resolved at the frequencies most commonly used. The geographical variation in texture shows up on the sonograph as a different shade (tone) of grey or black, the shade darkening with increasing roughness of the floor. In viewing each sonograph more attention should be paid to the tonal contrast within it rather than attempting to compare the shades on different sonographs, which will vary from figure to figure because of the operation of such factors as gain setting and tilt angle.

Echoes can sometimes be obtained from the sea floor at a slightly greater range than that allowed for on the recorder that is used (page 6), although these echoes are not usually a problem. An example of this effect is given in Fig. 89. More objectionable is an effect associated with a single frequency version of the dual side-scan system, where echoes from the sea floor to one side of a ship are picked up by the transducer facing the sea floor on the other side of the ship. This 'cross talk' can be confusing (Fig. 156 and 160).

The explanatory diagrams

Each sonograph is accompanied by an explanatory diagram and a brief description. In general, these diagrams show the trends of the main features and point out less obvious ones which might otherwise be overlooked. The *direction of sediment transport* on the floor is indicated by a thin arrow where necessary. This can sometimes be inferred from profiles of sand waves that are visible on the sonograph, but for others is only known from associated studies of the sediments and water movements. In a tidal environment the arrow shows the net direction of sediment transport, although material will be moving to a lesser extent to either side of the main direction, particularly when the tidal ellipse is broad. Some of the explanatory diagrams are drawn in true or nearly true plan view (the same scales in both dimensions) in order to show the real angular relationship of the different features that are present: in this case one arrow suffices to show the net sediment transport direction. On other diagrams, which are not a true plan view, the sediment transport arrow is drawn with respect to sand waves or sand ribbons, whichever is dominant.

The *length* and *width* measurements of the sea floor represented by the sonographs are shown on the outside of the explanatory diagrams. On many of them it has been necessary to indicate the near limit of the width measurement by a mark, as shown in Fig. 1. Care should be taken when using these distances to measure the size of features on the sonographs, as the width scale for

ground covered by the side lobes between the profile P and the near edge of the main beam is markedly nonlinear. For sonographs of the continental shelf (Fig. 2—87) the distance that is given applies only to ground illuminated by the main beam (where the scale is nearly linear): this value has been geometrically corrected for water depth, so that measurements may be made with greater certainty. Doubt about the gradients of the continental slopes make it impossible to apply a proper correction to the width measurement for these sonographs (Fig. 88—109) so that the mark has been placed at the profile P and the distance quoted is only corrected for water depth beneath the ship. Corrections were considered unnecessary for the sonographs in the second part of the book (Fig. 110—163); the distances given are the slant ranges as presented on the recorder. The *'illumination' direction* is shown by a broad arrow. The approximate *geographical co-ordinates* refer to the mid-points of the sonograph. The side-scan equipment that was used to obtain the data (see Table I) is only mentioned when it is not that fitted to the hull of R.R.S. "Discovery".

Presentation of sonographs

The primary division of the sonographs is into those showing the geology of the sea floor and those with other patterns. The geological section is arranged in the order: "Continental Shelf", "Upper Continental Slope" and "Deep Sea Floor". These geographical groups are then sub-divided into rocks, sediments and relief, but it must be emphasised that the aim is to give sonograph interpretations rather than to show the genetic relationships between different features. As it is not possible to obtain both high resolution and long range with the same equipment, the direct comparison of the sonographs obtained with equipment for either extreme might seem to denigrate the longer range equipment in terms of the detail shown. Accordingly, the sonographs obtained with short range gear are included under the headings "Continental Shelf" and "Upper Continental Slope", while the ones included under the heading "Deep Sea Floor" are limited to those obtained with GLORIA, the long-range equipment. Unfortunately, as GLORIA is such a relatively new development, there is a restriction on the choice of suitable sonographs, so that few are illustrated. The remaining section of the book is concerned with a wide variety of important non-geological effects which must be identified if they are not to be wrongly attributed to the geology of the sea floor. Wherever possible each pattern is shown separately. Dull, featureless floors, although of widespread occurrence, do

TABLE I

Parameters of side-scan sonar systems[1]

Originator	Equipment	Frequency (kilohertz)	Maximum recorder range (metres)	Pulse length (msec)	Main beam angles (degrees)		Mode of operation
					horizontal	vertical	
National Institute of Oceanography	GLORIA	6.5	22,000	12	2.7	10	towed
National Institute of Oceanography	Asdic (Mk I)	36	750	1	1.3	12	ship mounted
National Institute of Oceanography	Asdic (Mk II)	36	1,000	1	2.5	11	ship mounted
Géomécanique Compagnie Français	SOL-100	36.5	2 × 1,500	1	2	20	towed
Kelvin Hughes (Smiths Industries) Ltd.	Towed Surveying Asdic	48	1,500	2	1.6	7.5	towed
Kelvin Hughes (Smiths Industries) Ltd.	Transit Sonar	48	550	1	1.7	51	ship mounted
E.G. and G. International	Dual Side-scan Sonar	120	2 × 300	0.1	1	20	towed
Ministry of Defence (Naval)	ARL Electronic Sector Scanning Sonar	300	365	0.1	0.3	7.5	ship mounted

[1] The details are for equipment used to obtain the sonographs illustrated in the book. Range, pulse length, and main beam angles may be varied on certain equipment.

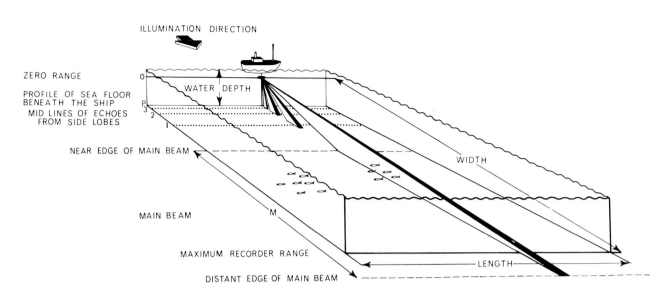

ILLUMINATION DIRECTION

ZERO RANGE

PROFILE OF SEA FLOOR
BENEATH THE SHIP
MID LINES OF ECHOES
FROM SIDE LOBES

WATER DEPTH

NEAR EDGE OF MAIN BEAM

WIDTH

MAIN BEAM

MAXIMUM RECORDER RANGE

LENGTH

DISTANT EDGE OF MAIN BEAM

Fig. 1. The *schematic diagram* shows the main sound beam and the side lobes that are used in the N.I.O. side-scan sonar system, together with the surface of the sea floor that is examined as the ship moves forward. The large arrow on this, and on all the *explanatory diagrams*, shows the direction in which the sound is sent out from a transducer beneath the ship. Other symbols are explained on pages 4 and 8. The two diagrams on this page should be compared with the *sonograph* on the opposite page. The white bands lying parallel with the ship's track are caused by the minimal sound energy reaching the floor between the sound beams.

Differences in tone on a sonograph of the sea floor are produced by relief or textural contrasts. However, it must be remembered that a particular slope or texture will not necessarily appear as the same tone on all sonographs. Some sea floor patterns are overprinted with additional patterns due to marine life, sea effects and man-made objects.

Explanatory Diagram

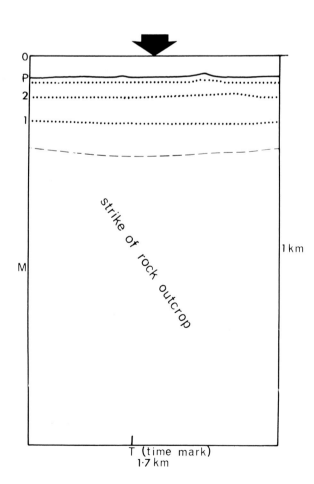

not need to be illustrated here.

Most of the sonographs are oriented with the depth profile P in a natural position at the top of the page. Because of this the 'illumination' direction for most of them is towards the bottom of the page, with any shadows that may be present falling towards the eyes of the reader. Unfortunately, these shadows are white rather than black and thus the relief does not appear realistic (Fig. 2). Of course, this difficulty can be partly overcome by viewing the sonograph upside down, to make the black areas become apparent shadows falling towards the viewer. However, it is more convenient to obtain black shadows that bring out the true relief by making use of a negative print (Fig. 3). *Negative prints* are indicated by the *thick black border* around their corresponding explanatory diagram. Ideally, perhaps, all sonographs should be shown in such a manner in order to make them directly comparable. However, this has not been done because only some rocky relief appears to be enhanced, and because it would conflict with the normal presentation of almost all other available equipment.

The full width of the original sonograph is only shown where it is an advantage to be able to see the relief revealed by the side lobes. In most cases the bands of echoes from these have been omitted, and the information that they provided is included in the figure captions. A few of the sonographs represent an almost true plan view, so that the true shape and orientation of features is apparent, but the great majority (in common with records obtained with most other existing side-scan equipment) have a somewhat exaggerated width scale. This, while causing a loss of true shape and orientation, can have some advantages, as with the vertical exaggeration of echo-sounder or sub-bottom reflection profiler records. Information along the ship's course is, in effect, compressed by a factor of up to about seven times. This shows more of the sea floor at a time than is possible on a true plan presentation, so that the considerable variation in form and composition can be more readily appreciated. The sonographs shown in this book have a rather better shape for illustration purposes than the equivalent long and extremely narrow true plan views. Also, the width exaggeration makes shadows more dramatic by lengthening them (since these are always presented at right angles to the ship's track, and so extend in the width direction of the sonograph). A further aspect is that elongate shapes other than those parallel or nearly parallel to the ship's track will have their linearity exaggerated and so made more obvious; whereas the apparent angular changes in trend will be greater than their true angular change when parallel or nearly parallel to the ship's track, and less than true when approaching right angles to the ship's track. Thus, the advantages and disadvantages are about equal. The ratio of width to length scales, the *width exaggeration* (W.E.) is generally given in the figure description.

I. GEOLOGICAL SONOGRAPHS

Sedimentary Rock

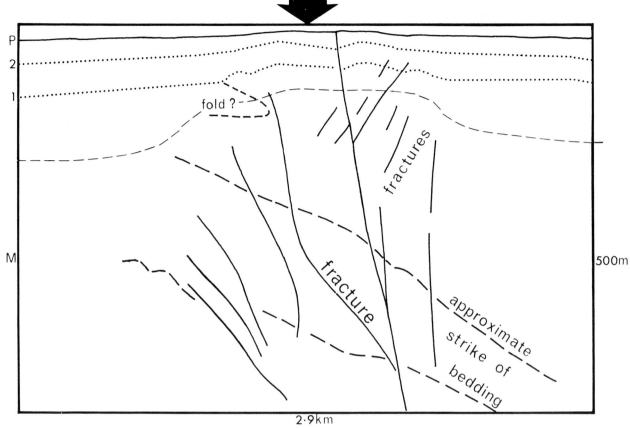

2·9km

Fig. 2 and 3. A ridge of Palaeozoic strata 3.2 km seaward of rocks of Carboniferous (Culm) age at Hartland Point, England. Individual beds are difficult to follow because of the presence of fractures and possibly some tight folds, also. The floor on either side is flat with a partial veneer of sand. Fig. 2 (upper sonograph) is a print of the original sonograph, with white shadows falling towards the viewer, while Fig. 3 (lower sonograph) is a negative print of the same ground. In Fig. 3 the eye can see unambiguously that the white areas are ridges and the black areas are shadows, because of the resemblance with familiar scenes on land. If this page is viewed upside down it is Fig. 2, with the shadows (white), now falling away from the observer, which appears to have the more obvious relief.

51°02.5′N   04°33′W
W.E. X 2.5

Continental shelf, Celtic Sea.
Water depth over ridge about 26 m.
Water depth around shoal about 35–42 m.

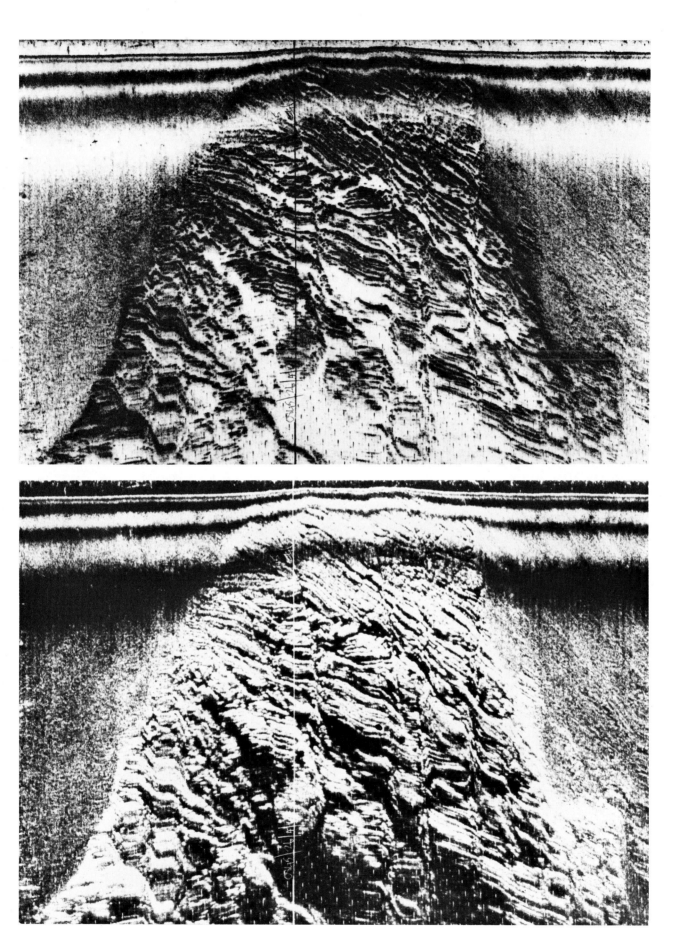

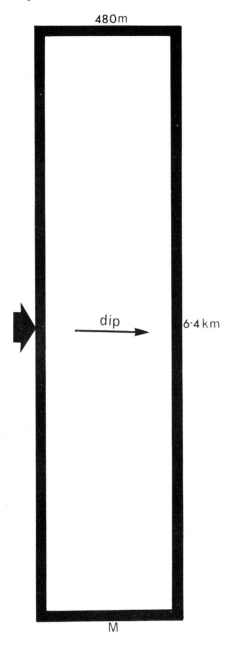

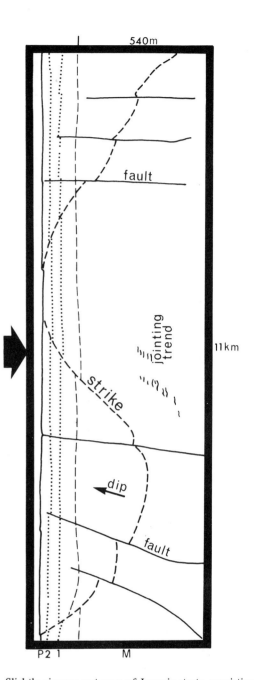

Fig. 4. Layers of relatively hard rock, of Jurassic age, that can be followed continuously for 6.4 km on the figure and up to 12 km on the adjacent parts of the sonograph. The ridges stand up to 5 m above the intervening softer layers, which have been etched out by tidal scour. The beds are known to dip towards the right of the figure. Shadows show up black on this figure.

Continental shelf, English Channel.   50°14′N 01°34′W
Water depth about 50 m.   W.E. X 3.5

Fig. 5. Slightly sinuous outcrops of Jurassic strata, consisting of alternate hard and soft layers, the latter etched out by tidal scour leaving the hard layers standing proud as ridges up to 10 m high. Continuity of the beds is fairly obvious despite their interruption by a series of small, almost parallel faults traceable for up to 1 km on the figure, and by smaller features thought to be joints. Note that for any bed the (black) shadow lengthens with increasing distance from the ship, as the angle of incidence of the sound decreases. In spite of this effect it can be seen that some strike ridges are taller than others, as shown by their relatively wide shadows. Some of these can be traced to steps in *P,2* and *1*, indicating that the beds dip towards the left of the figure.

Continental shelf, English Channel.   50°16′N 01°34′W
Water depth about 55 m   W.E. X 4.5

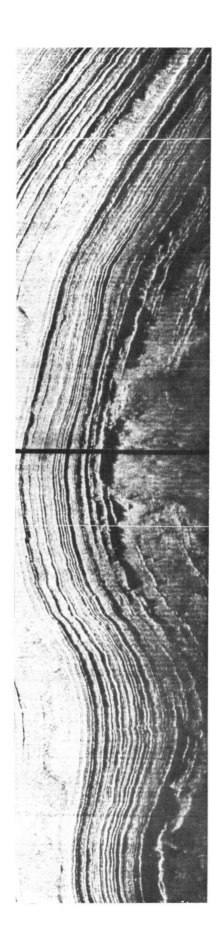

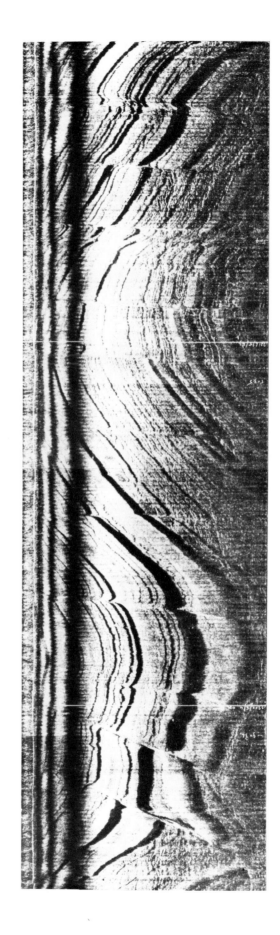

13

Sedimentary Rock

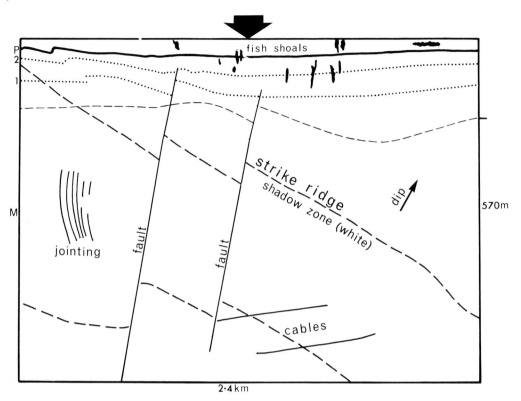

Fig. 6. Small, parallel dip faults displace rocks of Lower Jurassic (Liassic) age. Groups of soft layers of rock have been scoured into troughs, leaving three main groups of hard layers standing proud as ridges which cast long shadows (white zones).

Continental shelf, Bristol Channel.
Water depth up to about 30 m.

51°17′N 03°48′W
W.E. × 2.5

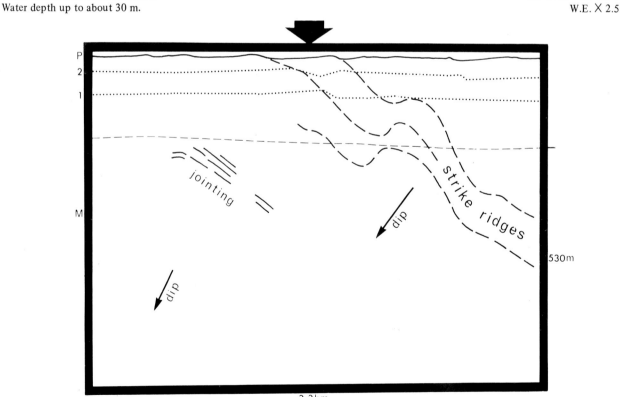

Fig. 7. Bold outcrops of Lower Jurassic strata. The harder layers stand up as ridges, their scarp slopes being shown as white lines facing the top of the page, with very gentle dip towards the bottom of the page. Some ridges, and the intervening troughs, floored by softer rocks, can be followed through the side lobes P, 2 and 1, to the edge of the main beam, M. The sinuosity of the outcrops and the accentuation of the relief probably results from a combination of erosion by ice and rivers during periods of Quaternary low sea level, and subsequently by the scouring action of sand moved by tidal currents. Shadows show up black on this figure.

Continental shelf, Bristol Channel.
Water depth about 30 m.

51°19′N 03°42′W
W.E. × 2.5

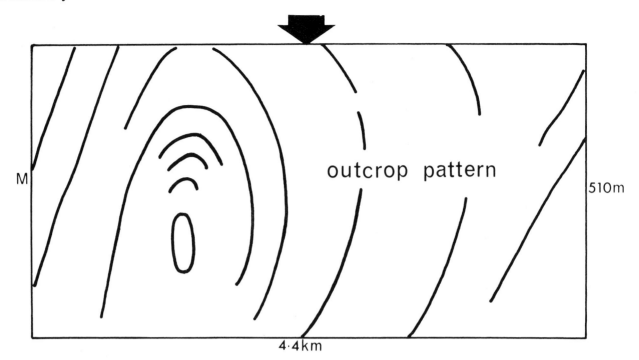

Fig. 8. Low, just discernible, outcrops of thin bedded strata of Mesozoic age, defining a small structure in the general region of salt domes off northeastern England.

Continental shelf, North Sea.
Water depth about 40 m.

54°39′N 00°52′W
W.E. × 4

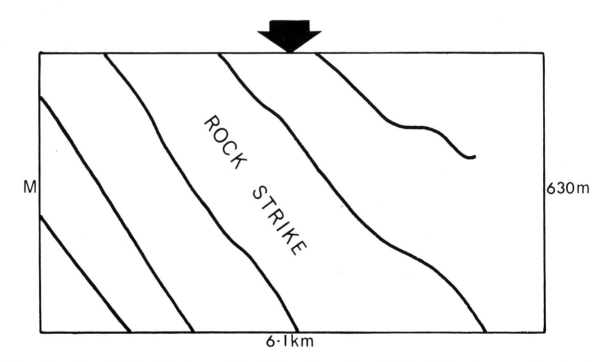

Fig. 9. Low outcrops (wavy black lines) of rock separated by flat floor of loose sediment (light tone). This rock pattern might perhaps be confused with sand ribbons.

Continental shelf, north of Spain.
Water depth about 140 m.

43°39′N 05°09′W
W.E. × 5.5

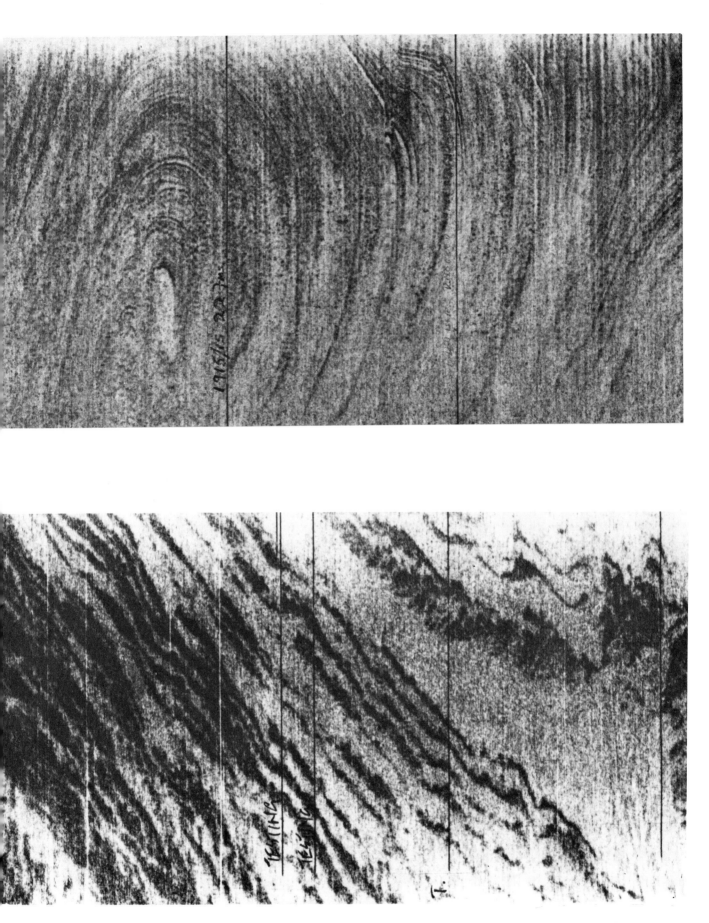

Sedimentary Rock

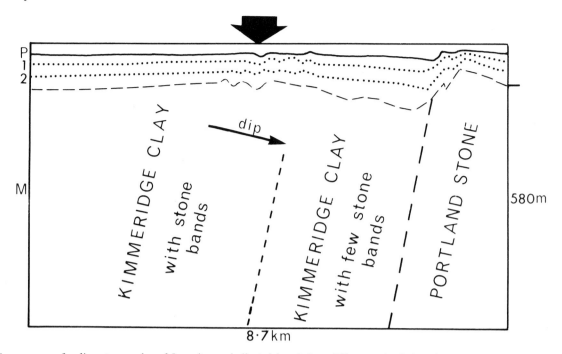

Fig. 10. A sequence of sedimentary rocks, of Jurassic age, indicated by obvious differences in their resistance to erosion. The softest rocks in the middle of the sequence find surface expression as a trough, which separates older beds with hard layers (on the left) from the ridge of Portland Stone (on the right). The scarp edges of the beds appear as dark lines facing towards the left.

Continental shelf, English Channel.
Water depth about 17–42 m.

50°32.5'N   02°08.5'W
W.E. × 7

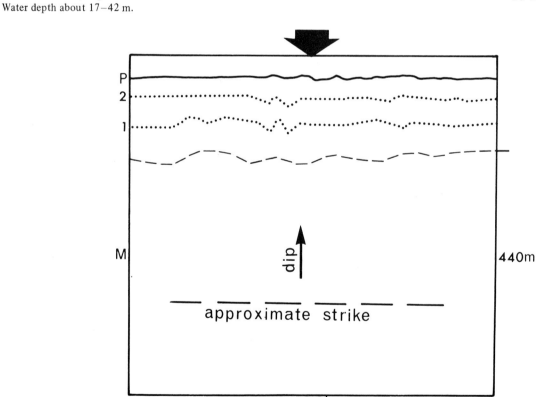

Fig. 11. A rough floor (shown at *1, 2* and *P*) giving rise to irregular shadows (white) whose alignment indicates the approximate strike of the outcropping layers of rock, probably Cretaceous.

Continental shelf, English Channel.
Water depth about 60 m.

50°16'N   01°17'W
W.E. × 3

18

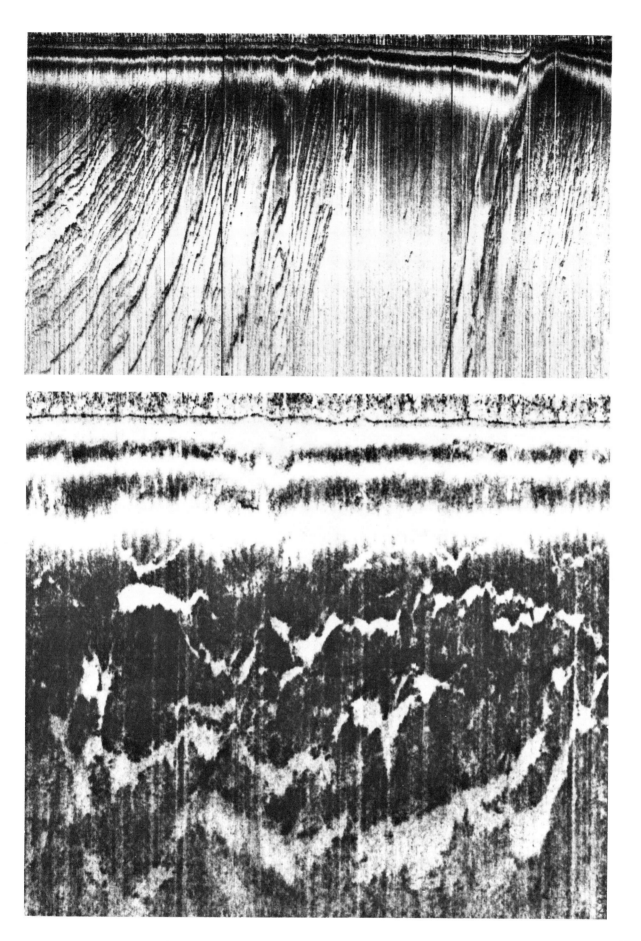

Sedimentary Rock

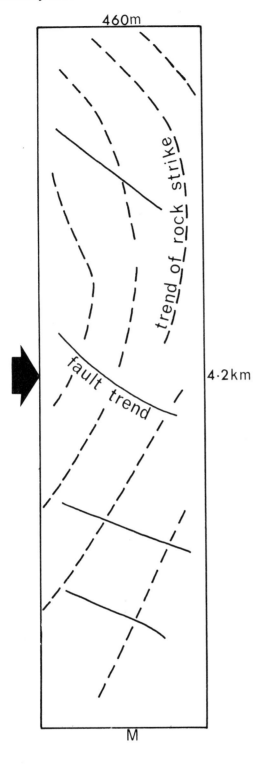

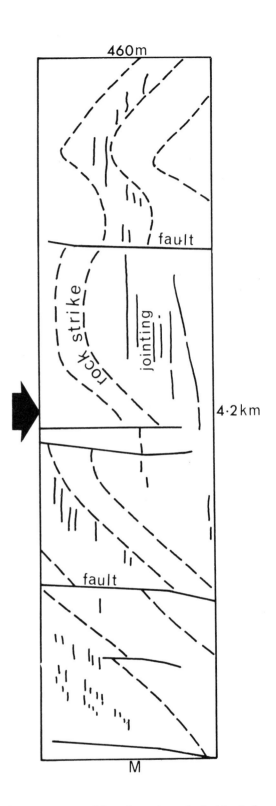

Fig. 12. A parallel group of faults, up to 400 m apart, that maintain their trend and direction of downthrow in spite of the changing strike of the Jurassic strata. Ground in shadow appears white.

Continental shelf, English Channel.   50°14'N 01°37'W
Water depth about 57 m.   W.E. × 2.5

Fig. 13. Low outcrops of Jurassic strata emphasised by shadows (white). These layers are displaced by faults, normal to which there is a particularly well-developed joint pattern that is most obvious at the edge of the beds.

Continental shelf, English Channel.   50°15'N 01°32'W
Water depth about 55 m.   W.E. × 2.5

21

Sedimentary Rock

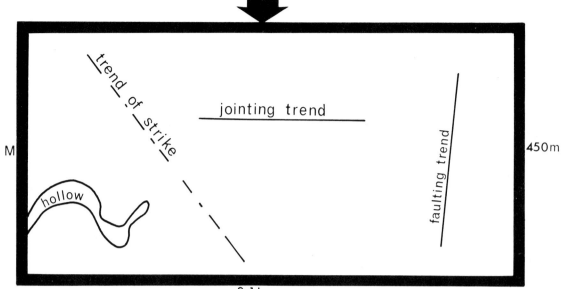

Fig. 14. Almost straight ridges of relatively resistant Jurassic strata displaced by small parallel faults. There is an obvious secondary pattern of linear relief features about 10 m apart, which is thought to indicate joints etched out by marine erosion in all but the more resistant layers. The origin of the sinuous hollow at the bottom left side of the sonograph is unknown. The shadows (black) lengthen from top to bottom of the sonograph as the angle of incidence of the sound decreases.

Continental shelf, English Channel.   50°15′N   01°29′W
Water depth about 51 m.   W.E. × 2

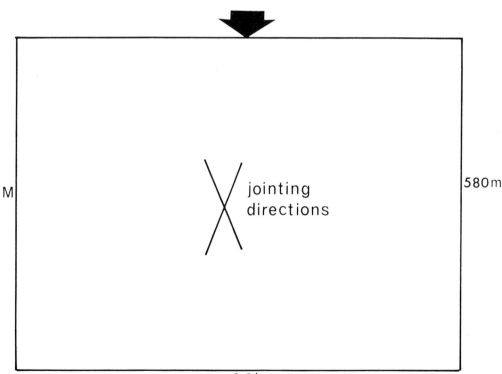

Fig. 15. Two sets of joints in limestones of Upper Jurassic (Purbeck) age. They are best shown by the zig-zag, broad black lines representing the edges of the relatively thick layers, with shadows (white) behind some of them.

Continental shelf, English Channel.   50°28′N   02°23′W
Water depth about 45 m.   W.E. × 3

22

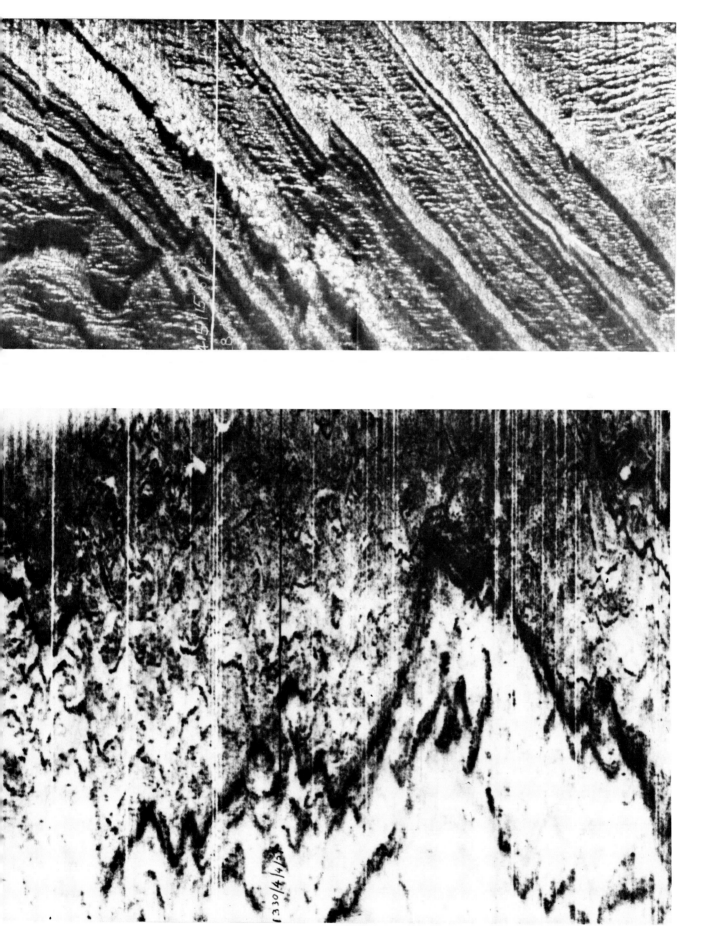

Sedimentary Rock

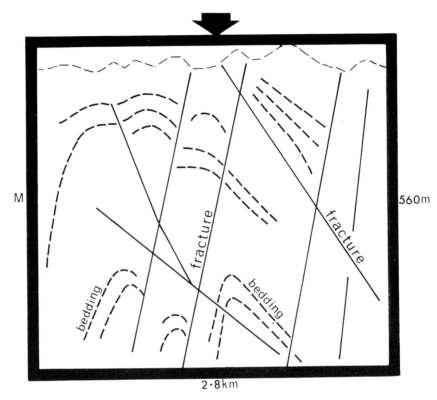

Fig. 16. Highly fractured and folded strata, probably low grade slates of Devonian or Carboniferous age. Shadows are black on this figure.

Continental shelf, Bristol Channel.   51°14′N   04°42′W
Water depth about 40 m.   W.E. × 5

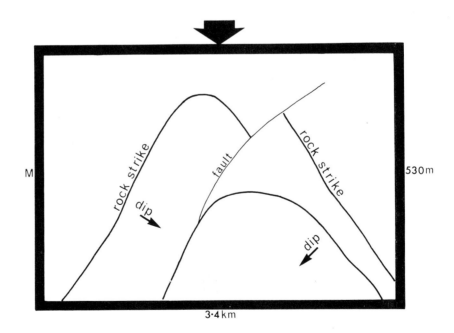

Fig. 17. A small fold displaced by a curving fault in well-exposed rocks of Upper Jurassic age. The edges of relatively hard layers show up as narrow, white lines, the dip slopes as broad dark bands. The intervening ground of relatively soft rock has a flat surface. Shadows appear black on this figure.

Continental shelf, English Channel.   50°14′N   01°37′W
Water depth about 55 m.   W.E. × 4

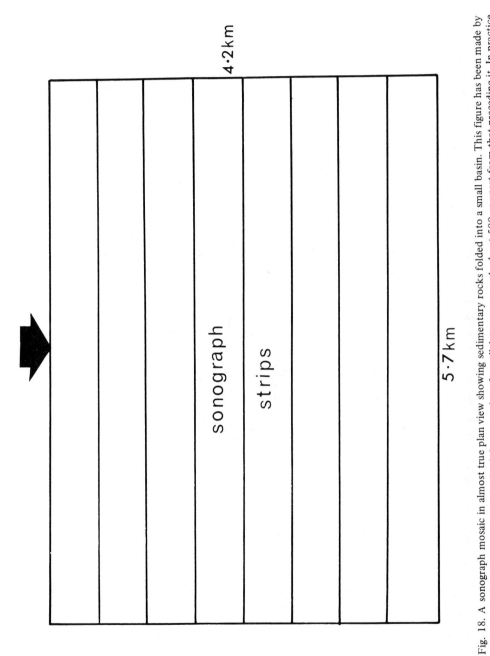

Fig. 18. A sonograph mosaic in almost true plan view showing sedimentary rocks folded into a small basin. This figure has been made by putting together the sonographs obtained from a series of parallel traverses, each about 500 m apart from that preceding it. In practice the difficulties of navigation and moving over the ground at constant speed make it difficult to match the outcrops from traverse to traverse exactly, without some processing of the data, although in this example those errors do not mask the structure.

43°29′N   01°39′W
W.E. × 1

By courtesy of Institut Français du Pétrole.
Géomécanique SOL-100.
Water depth about 15-80 m.

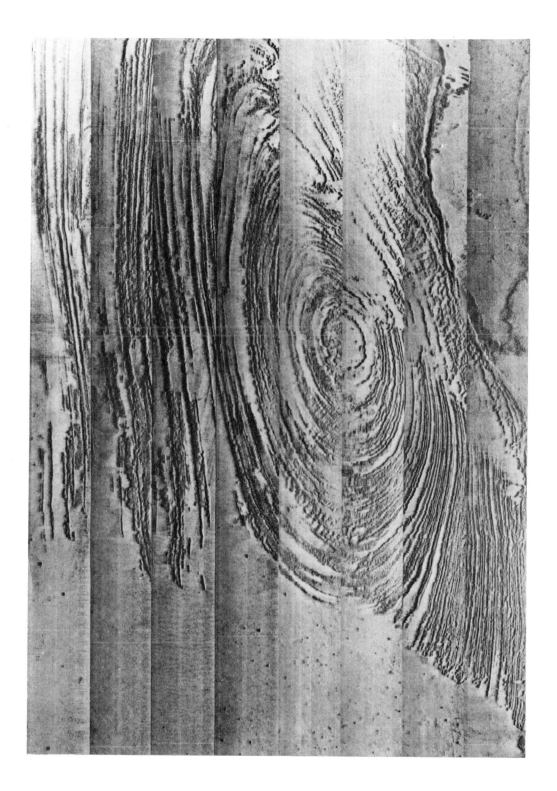

Igneous or Metamorphic Rock

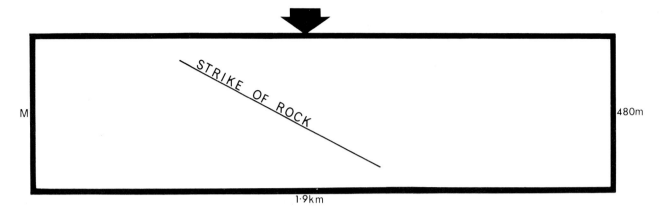

Fig. 19. Near true-plan view of outcrops of massive rock, located between the metamorphic rocks on the islands of Coll and Tiree and the basalts of Mull. Shadows in this figure appear black.

Continental shelf, Scotland.
Maximum water depth about 45 m.
Towed Surveying Asdic.

56°30.5′N 06°32.5′W
W.E. × 1
By courtesy of the Geophysics Group, Bath University.

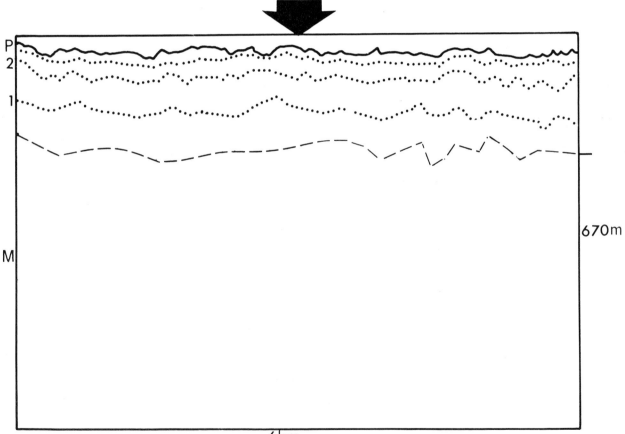

Fig. 20. Rough rock floor with rather indefinite trends. It is analogous in outcrop pattern to that of the glaciated Precambrian gneisses occurring 40 km away in the Outer Hebrides, Scotland. Between the outcrops there is a partial sediment infill.

Continental shelf, west of Scotland.
Water depth about 130 m.

57°23′N 08°10′W
W.E. × 4

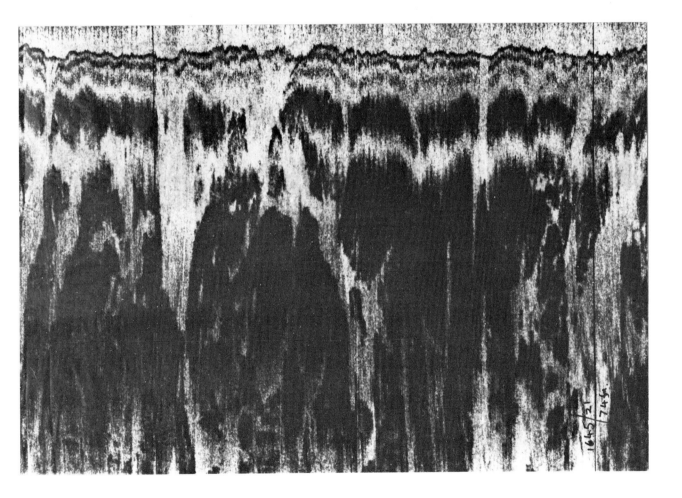

Igneous or Metamorphic Rock

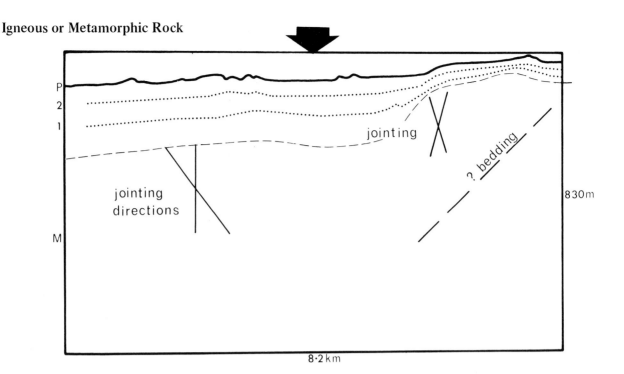

Fig. 21. Two sets of joints interrupting rock outcrops (up to about 20 m high), whose pattern strongly resembles the terrain of metamorphic and igneous rocks seen 10 km distant at the coast of northwestern Spain. The right hand outcrop may also have traces of bedding. Smooth sand floor occurs between the main outcrops and partially fills the enlarged lines of fracture.

Continental shelf, west of Spain.   43°00'N 09°23'W
Maximum water depth about 110 m.   W.E. × 5

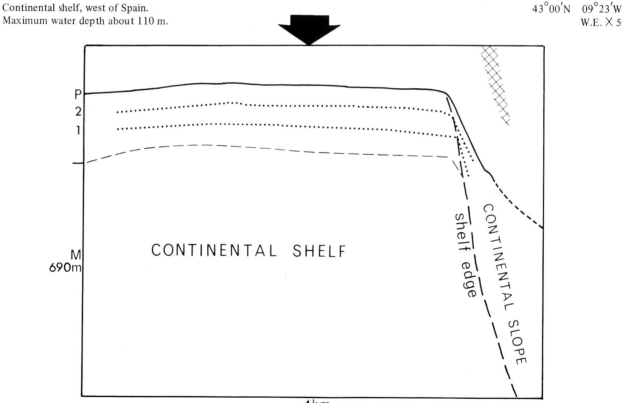

Fig. 22. The oceanward edge of the continental shelf showing rough ground, probably igneous rock, between which there is smooth floor of loose sediment. Strong reflections from the continental slope just beyond the bottom right of the figure are recorded in an apparent midwater position.

Continental shelf, west of Portugal.   39°33'N 09°32'W
Minimum water depth about 130 m.   W.E. × 2.5

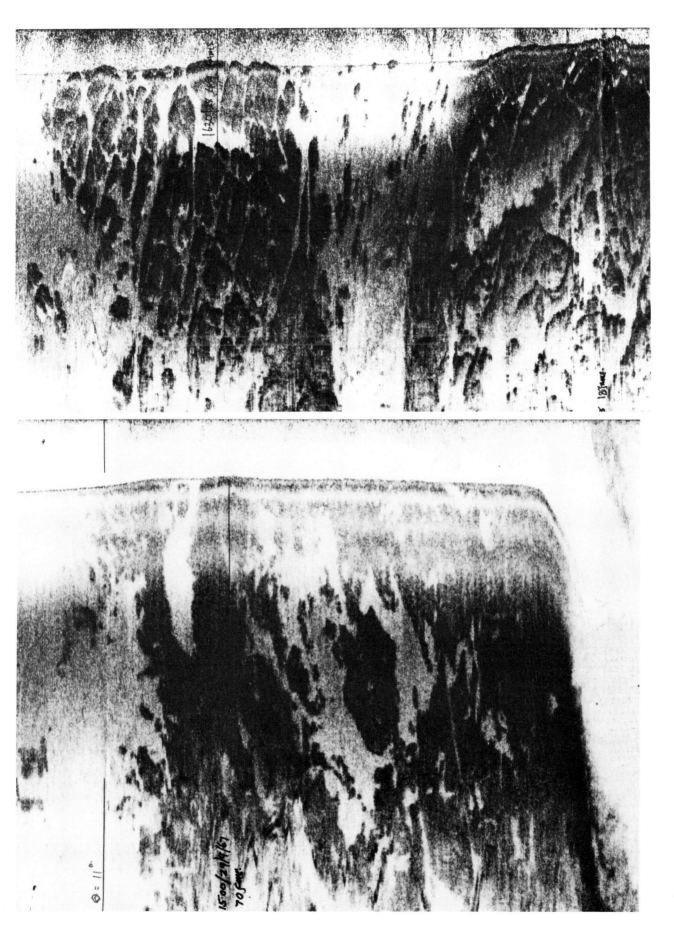

Igneous Rock

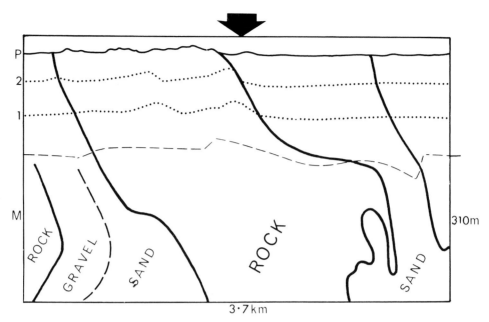

Fig. 23. Small outcrops of granite of Variscan age (the Haig Fras Granite), with characteristic blocky pattern, separated by linear areas of sediment infill indicating a major direction of fracturing. Some of the outcrops are high enough to cast shadows (white).

Continental shelf, Celtic Sea.   50°13'N 07°52'W
Depth about 90 m.   W.E. X 3.5

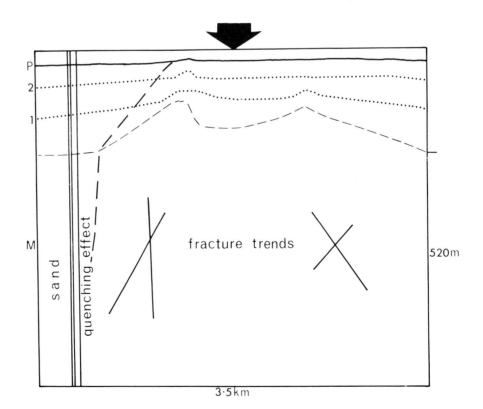

Fig. 24. Rock outcrops broken by fractures into relatively small blocks which cast shadows (white). The outcrop pattern is similar in appearance to that of some granites but in the present case may be due to some other hard rock, possibly a gneiss or a sandstone.

Continental shelf, English Channel.   50°12'N 04°48'W
Water depth about 45 to 80 m.   W.E. X 4.5

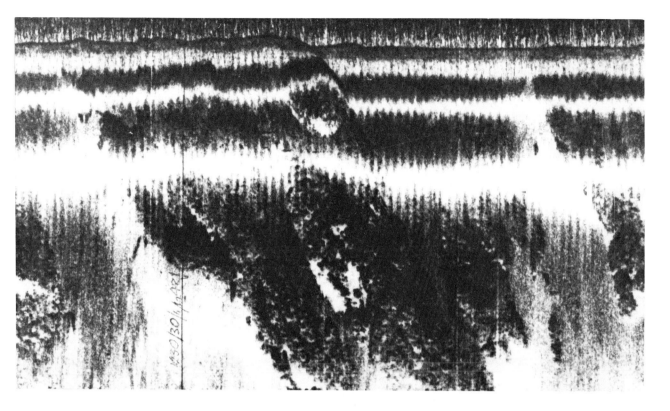

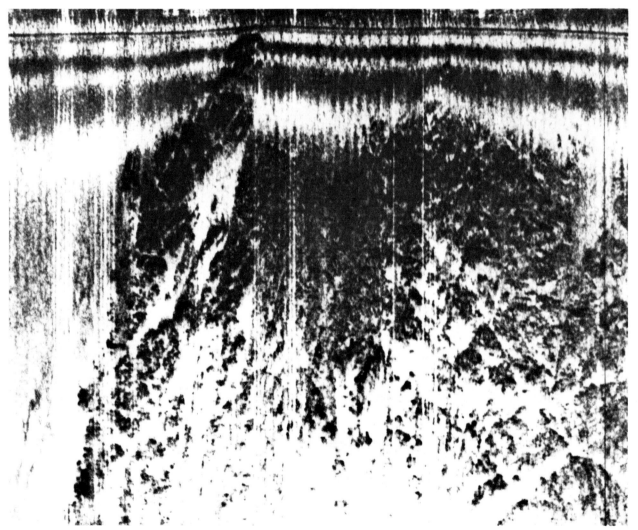

Igneous or Metamorphic Rock

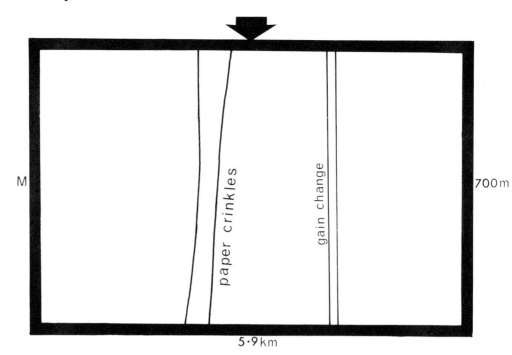

Fig. 25. A group of isolated rock outcrops with high relief (perhaps up to 30 m) casting shadows (black), surrounded by flat sand floor.

Continental shelf, northern entrance to the Irish Sea.
Maximum water depth about 130 m.

54°28'N 05°15'W
W.E. X 5.5

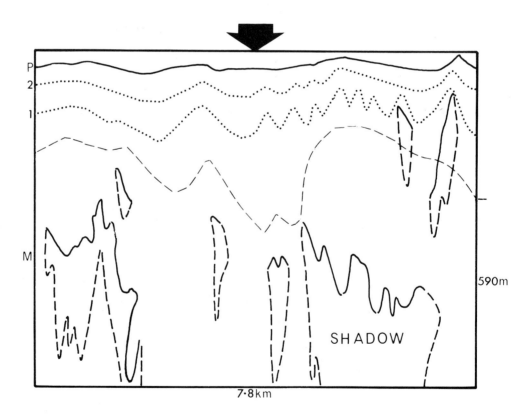

Fig. 26. Shadows (white) up to 400 m long, cast by ragged rock outcrops up to about 40 m high, in a region of glaciated relief. The floor between the outcrops consists of loose sediment.

Continental shelf, north of Ireland.
Water depth from about 66 m to about 125 m.

55°25.5'N 06°34'W
W.E. X 5.5

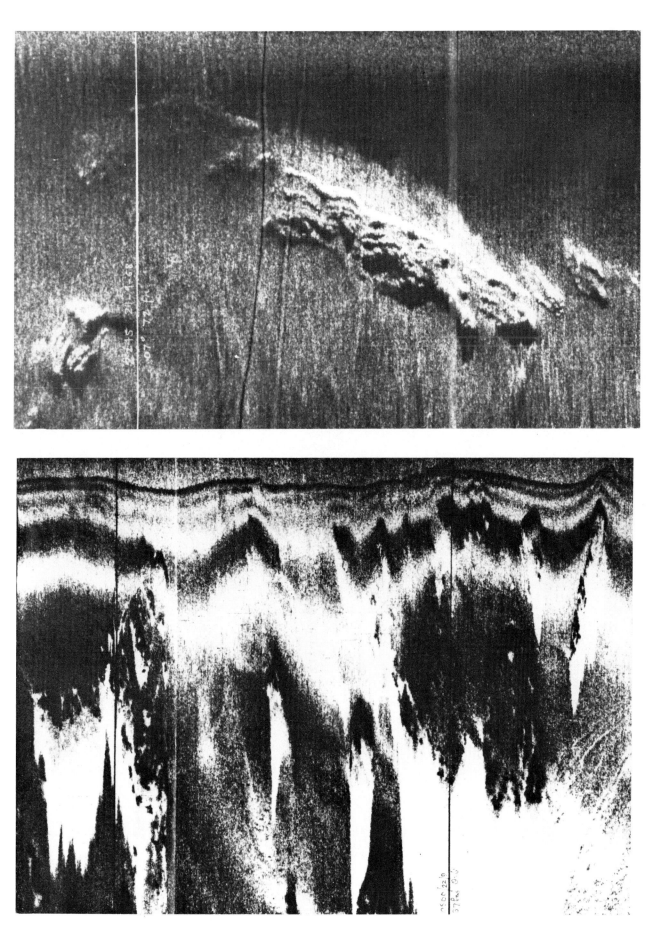

35

Sediment—Longitudinal Furrows

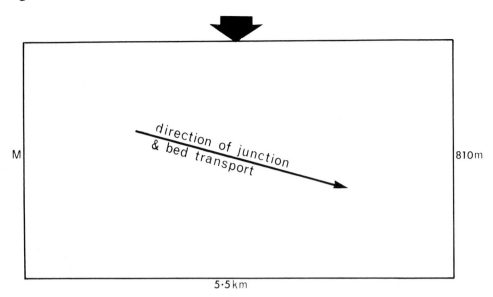

Fig. 27. Furrows more than 5 km long, 10 m wide and 1 m deep and up to about 60 m apart, which are notable for the well-defined branching pattern. The floor as a whole is almost flat and made of gravel. The furrows are elongated parallel with the strongest tidal currents and the direction of junction seems to agree with the direction of net transport of gravel. For each furrow the side facing the illumination direction provides a strong reflection and so appears as a black line, while the other side is in shadow and so appears as a white line. Some of the apparent sinuosity of the furrows may be due to ship's yaw.

Continental shelf, English Channel.   50°19′N   01°59′W
Water depth about 45 m.   W.E. × 3.5

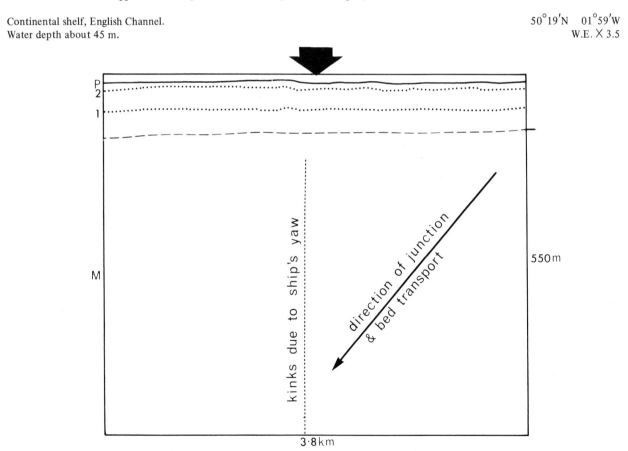

Fig. 28. A system of branching furrows, the direction of junction agreeing with the local net bed-transport direction. The profile shows some indications of their relief.

Continental shelf, English Channel.   50°27′N   00°58′W
Water depth about 30 m.   W.E. × 4.5
Kelvin Hughes Towed Surveying Asdic.   By courtesy of Geophysics Group, Bath University.

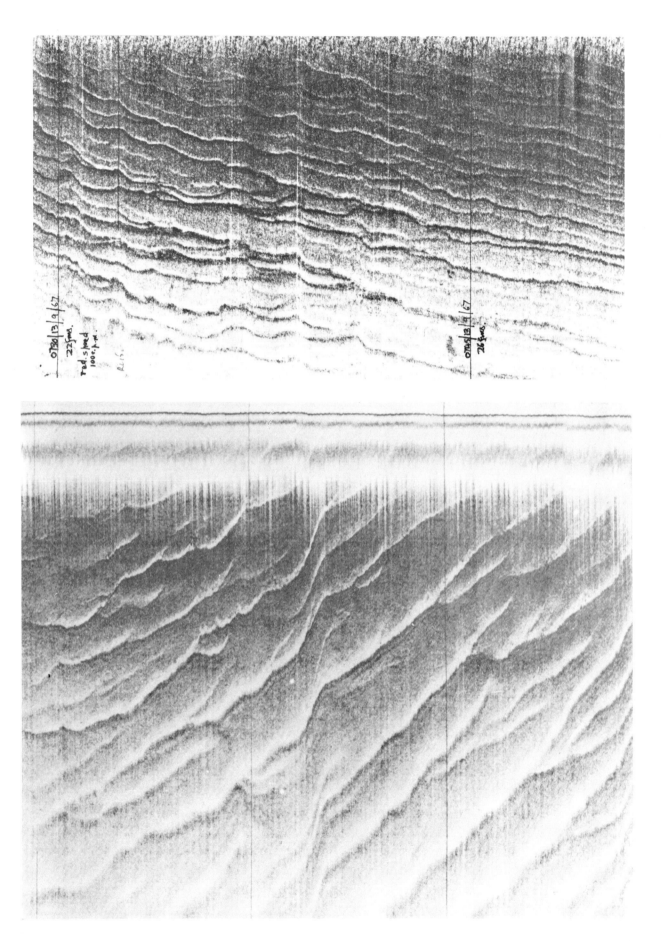

Sediment—Longitudinal Furrows

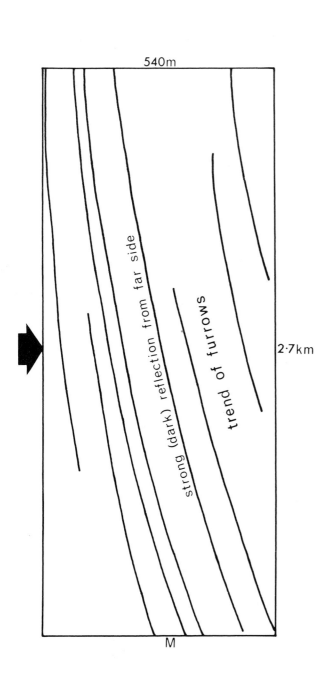

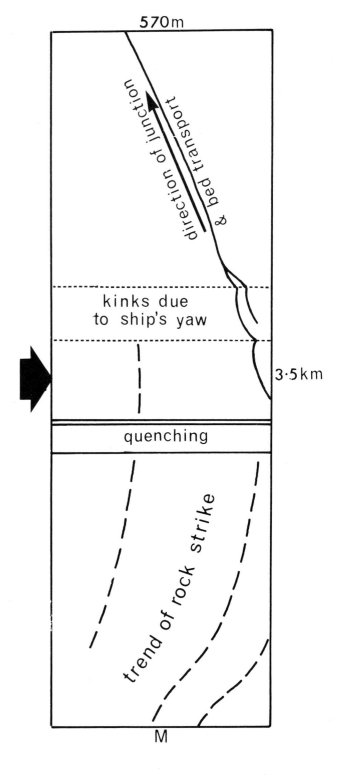

Fig. 29. Sharply defined longitudinal furrows up to 5 km long, 25 m wide and up to 1 m deep. They are notable for being almost straight and for having few branches compared with Fig. 27. They occur in a region of gravel but may be partly floored with sand. Their parallelism with the strong tidal currents suggests a causal relationship.

Continental shelf, Bristol Channel. 51°21.5′N 04°12′W
Water depth about 37 m. W.E. × 2

Fig. 30. Prominent furrows aligned across low rock outcrops. The furrows join down the sediment transport path. The side of each furrow facing the illumination direction gives a strong reflection and so appears as a black line, while the near side is in shadow and so appears in white. In contrast the rock ridges have the shadow on the side furthest from the illumination direction.

Continental shelf, Bristol Channel. 51°19.5′N 04°10.7′W
Water depth about 35 m. W.E. × 2

Sand Ribbons

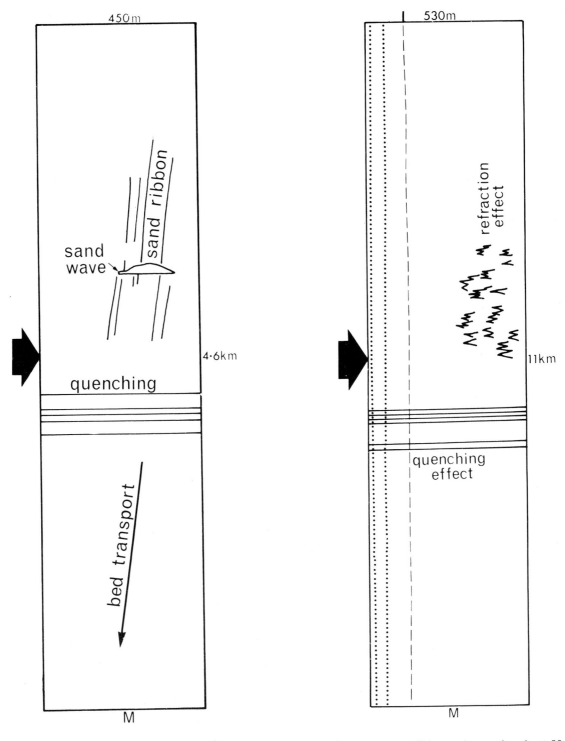

Fig. 31. Almost straight sand ribbons, traceable for more than 5 km, which show only small progressive changes in width and separation and whose variable pale tone suggests longitudinal and transverse differences in sand thickness. The pattern is interrupted by an isolated sand wave. The slight sinuosity is probably due to ship's yaw.

Continental shelf, Irish Sea. 52°28′N 05°54′W
Water depth about 70 m. W.E. × 2.5

Fig. 32. Sand ribbons (light tone) averaging about 90 m wide with clearly defined edges, and with sufficient thickness locally to cast small shadows. Some ribbons are remarkable for their curvature, wavy edges, and arrangement en échelon which results from their relatively small length to breadth ratio. These characteristics probably result from the location of the ribbons near to the channel between Lands End and the Scilly Isles, England, where ebb and flood tidal currents follow different paths and the tidal ellipse is rather broad.

Continental shelf, English Channel. 49°47′N 05°55′W
Water depth about 90 m. W.E. × 4

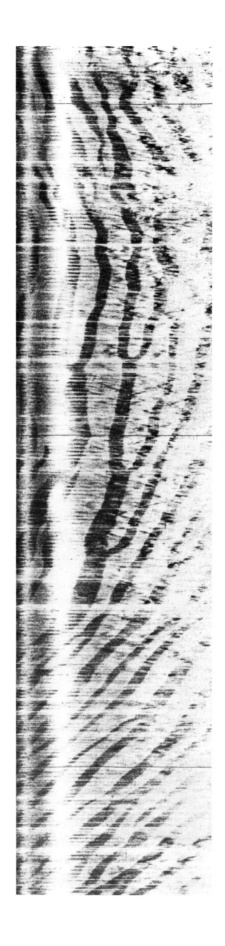

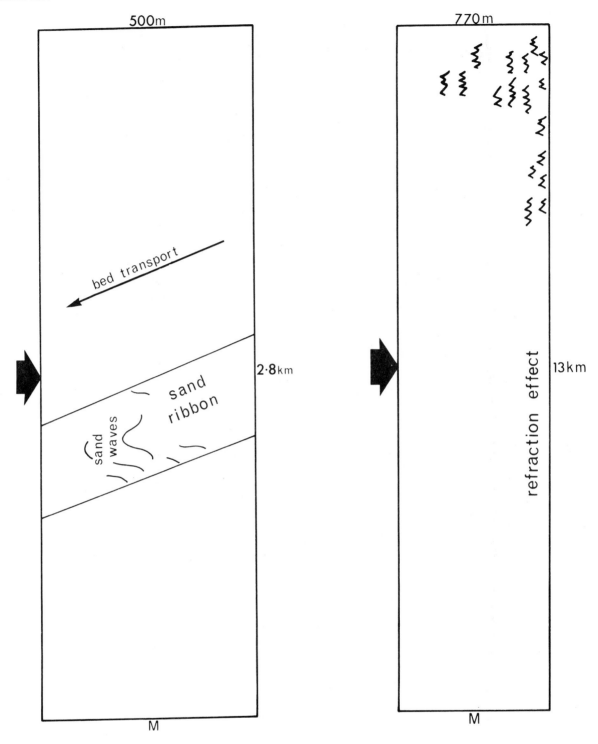

Fig. 33. An almost flat gravel floor (dark tone) with sand ribbons of variable width and separation. The patchy distribution of sand (light tone) within their confines suggests the presence of low, short-crested, irregular sand waves less than 1 metre high.

Continental shelf, English Channel.   50°11′N 02°33′W
Water depth about 59 m.   W.E. X 1.5

Fig. 34. Rhythmic groups of broad and narrow sand ribbons (light tone), the broadest having diffuse borders with the adjacent broad strips of coarser floor (dark tone). Some of the waviness of the ribbons results from ship's yaw.

Continental shelf, Celtic Sea.   49°34′N 06°09′W
Water depth about 100 m.   W.E. X 4

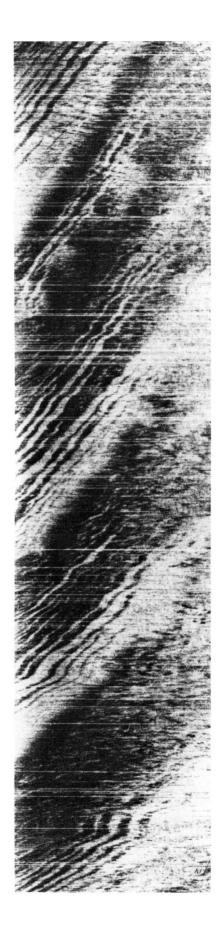

43

Sand Ribbons

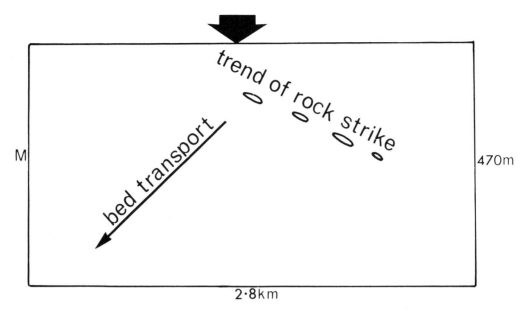

Fig. 35. Narrow and almost straight sand ribbons on a gravel and rock floor. The rock outcrops are low and isolated but nevertheless interfere with and serve as points of origin for some of the sand ribbons. An indication of the strike of the rock can be gained by viewing obliquely from the bottom right corner.

Continental shelf, Bristol Channel.   51°15′N   04°43′W
Water depth about 49 m.   W.E. X 3.5

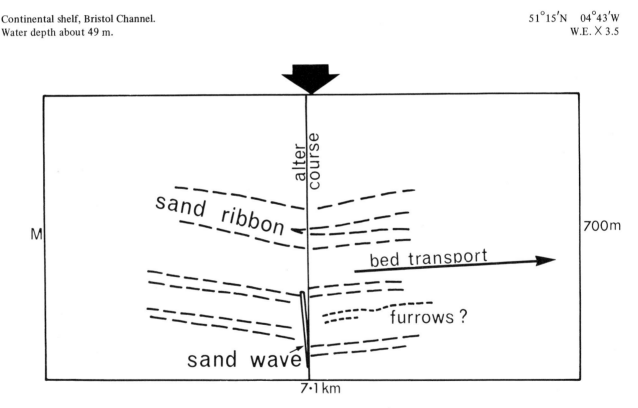

Fig. 36. Interpreted as showing the bifurcation of a sand ribbon in the direction of sediment transport and the interruption and lateral displacement of two other sand ribbons by an isolated sand wave. There are also faint indications of longitudinal furrows. The slight change in trend of the ribbons at about the middle of the sonograph is due to a course alteration.

Continental shelf, English Channel.   50°13′N 02°25′W
Water depth about 57 m.   W.E. X 5.5

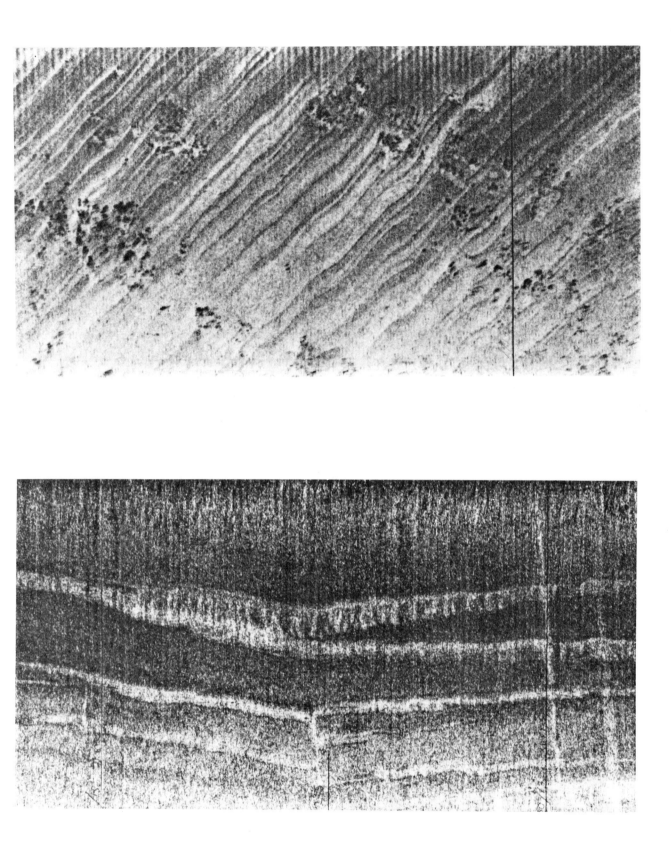

Sand Ribbons

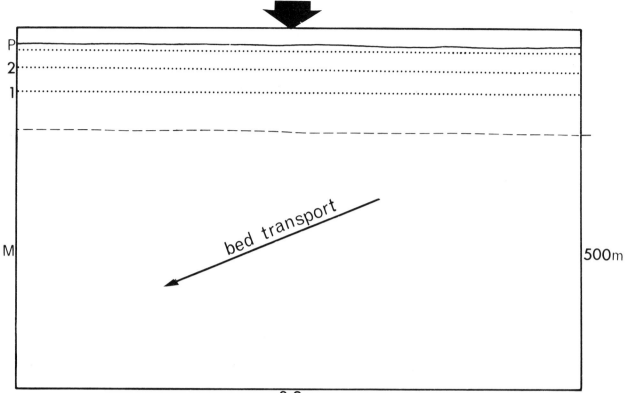

Fig. 37. Ribbons of sand (light tone) on a flat floor of coarser material (dark tone). These ribbons narrow progressively and bifurcate with increasing current strength (from right to left). This example is notable, in comparison with most others, in that the direction of bed-load transport is heading up the local tidal current velocity gradient. The regular pattern of dots particularly visible in the bottom right corner is due to echo-sounder interference.

Continental shelf, Bristol Channel.　　　　　　　　　　　　　　　　　　　　51°14.7′N　04°35.5′W
Water depth about 44 m.　　　　　　　　　　　　　　　　　　　　　　　　　　W.E. × 2.5

Sand Ribbons and Sand Waves

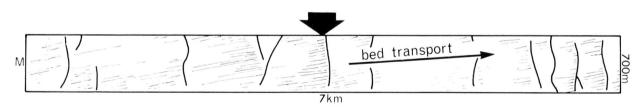

Fig. 38. Short, narrow sand ribbons parallel with the transport path and lying on the flat gravel floor between sinuous sand wave crests about 2 m high. This suggests loss of sand from the crests. Some of the sand ribbons are continuous between adjacent crests whilst others commence on the intervening floor. There is a tendency for them to broaden in the direction of net sand transport. The real angular relationship of the two sets of features is shown in the diagram at true plan view (W.E. × 1).

Continental shelf, English Channel.　　　　　　　　　　　　　　　　　　　　49°52′N　04°15′W
Water depth about 73 m.　　　　　　　　　　　　　　　　　　　　　　　　　　W.E. × 4.5

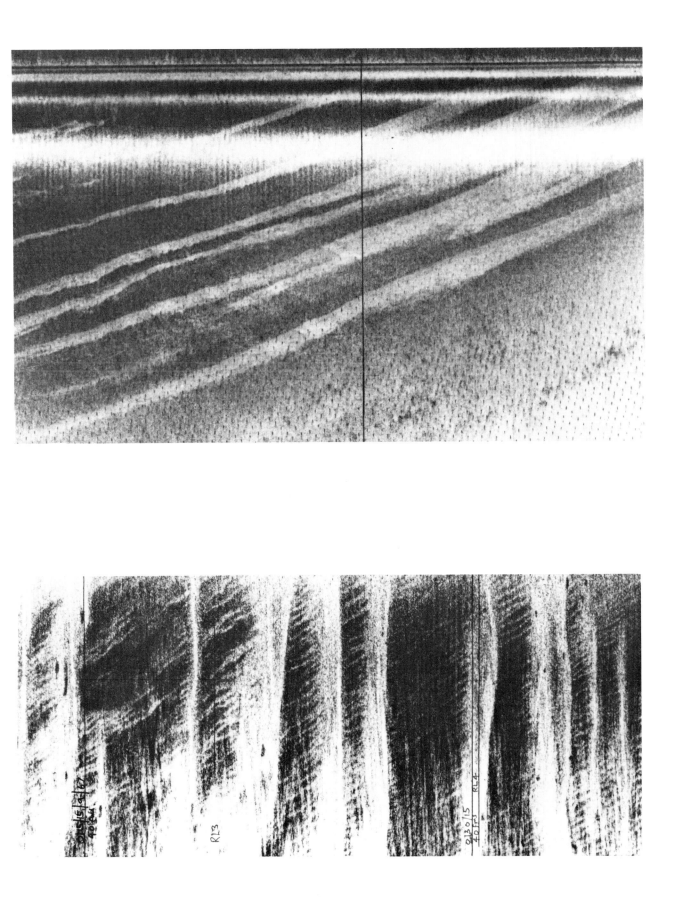

Sand Ribbons and Sand Waves

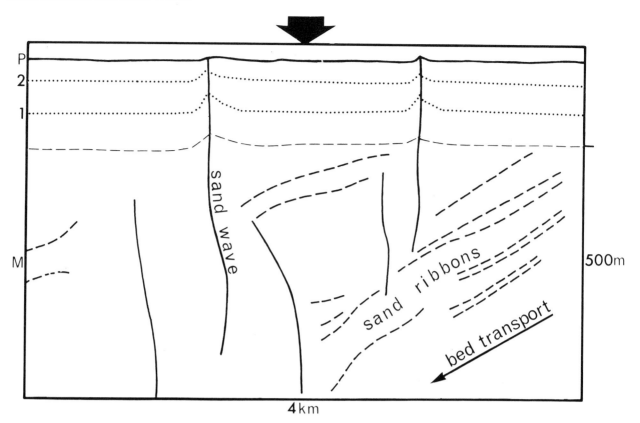

Fig. 39. Broad, short sand ribbons between sand wave crests, on a flat gravel floor. The sand ribbons are aligned almost normal to the sand wave crests, the angular discordance on the sonograph being due to the width exaggeration. The saw tooth effect prominent at the edges of the sound beams *M, 1* and *2* is due to a mechanical defect in the beam stabilisation system.

Continental shelf, English Channel.
Water depth about 55 m.

50°07′N 03°04′W
W.E. × 3.5

Sand Waves

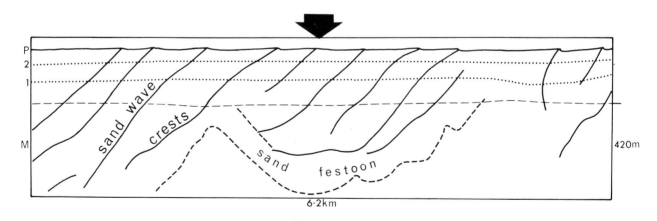

Fig. 40. Large sand waves with a wavelength of about 120 m, height about 6 m, and with asymmetrical cross-sectional profiles, separated from one another by a gravel floor. They are remarkable for the way in which some of the crests emanate from a broad but thin festoon of sand.

Continental shelf, English Channel.
Maximum water depth in troughs about 108 m.

48°24′N 05°26′W
Sonograph W.E. × 6
Diagram W.E. × 2

49

Sand Ribbons and Sand Waves

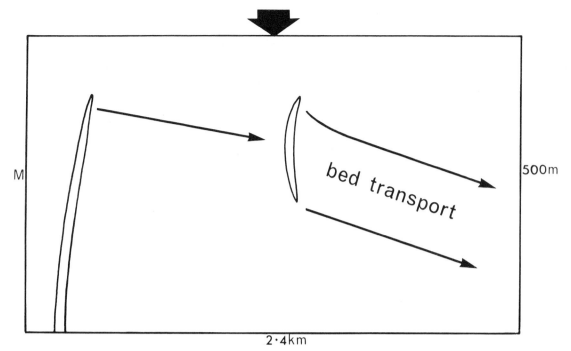

Fig. 41. Two isolated, slightly crescentic sand waves shedding sand from their wings in the form of diffuse ribbons (light tone) which widen slightly in the net transport direction. The regular pattern of dots is due to echo-sounder interference.

Continental shelf, Bristol Channel.
Water depth about 29 m.

51°24′N 04°02′W
W.E. × 3

Sand Ribbons

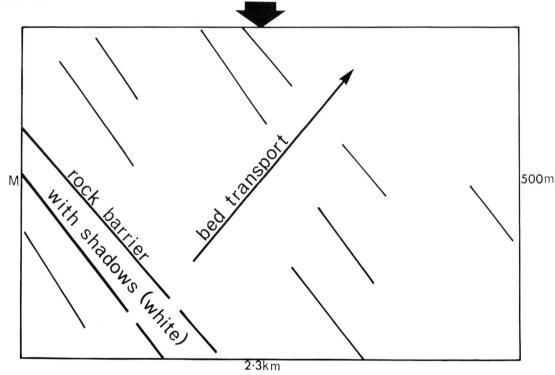

Fig. 42. A thin layer of sand (light tone) fanning away from gaps in a rock barrier, up to 20 m high, and passing over a series of lower rock ridges.

Continental shelf, Bristol Channel.
Water depth about 27 m.

51°14.5′N 04°13′W
W.E. × 4

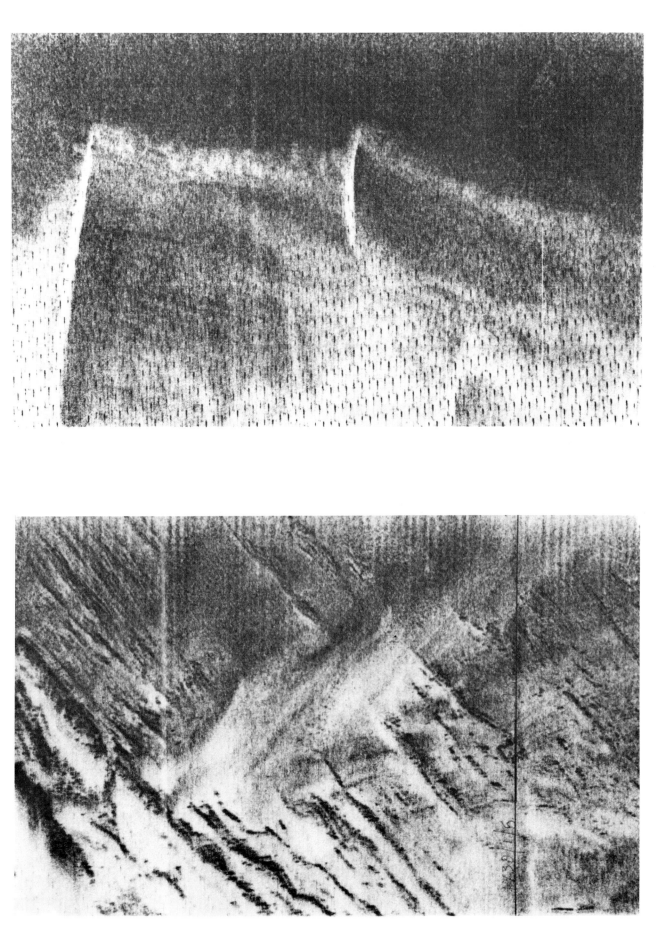

51

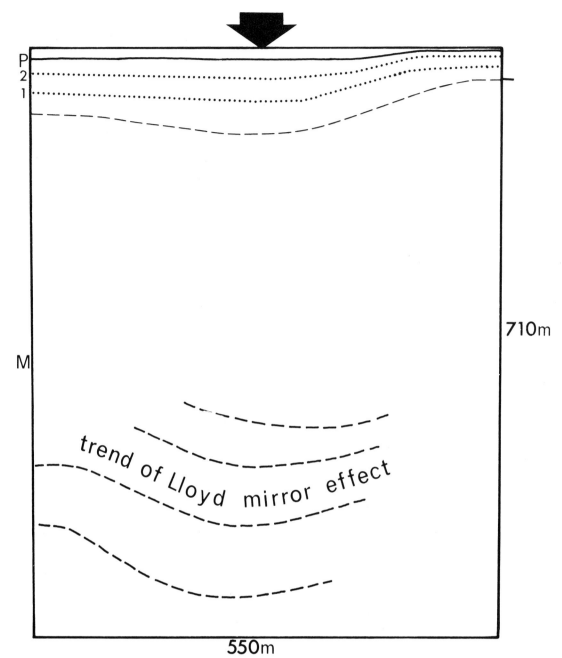

Fig. 43. A discrete train of straight, long-crested sand or gravel waves on the floor of a trough. The waves are about 10 m apart, and tend to interfinger and divide. The regular pattern of dots seen towards the bottom of the figure is due to echo-sounder interference, while the broad dark and light banding noticeable in the lower part of the sand wave train is a Lloyd Mirror effect.

Continental shelf, Bristol Channel.   51°15′N   03°46′W
Water depth about 8 to 18 m.   W.E. × 1

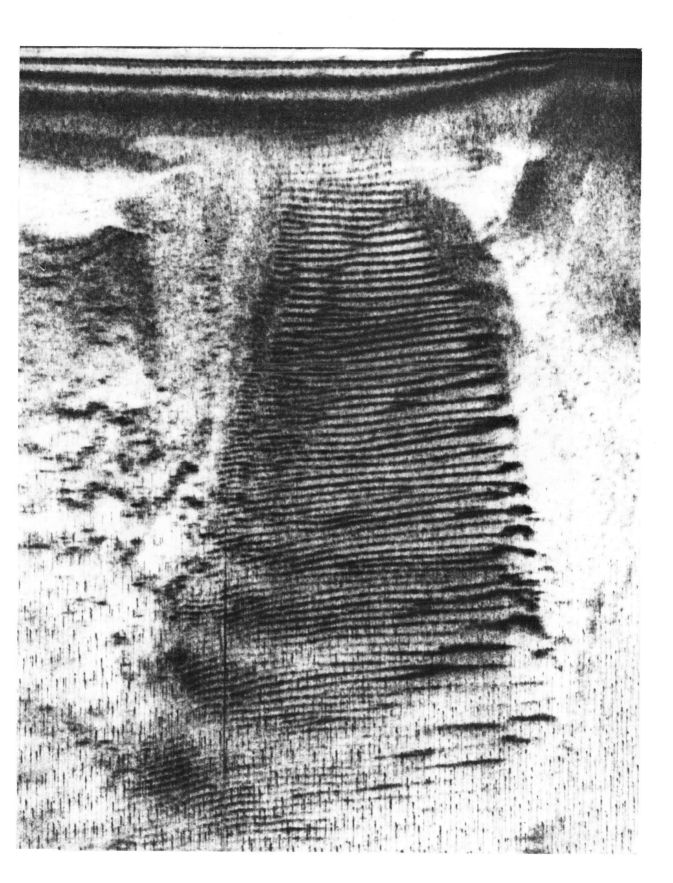

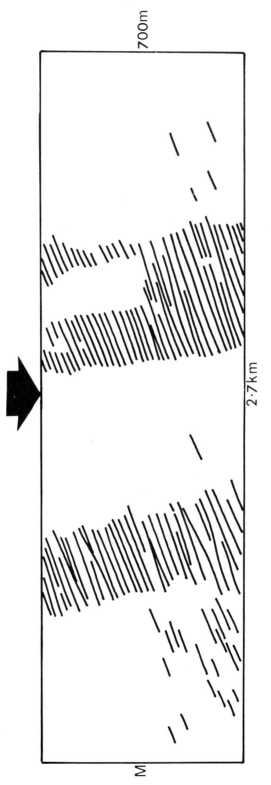

700m

2·7km

M

Fig. 44. Two trains of small gravel waves with indications of more on the surrounding flat floor. Their wavelength is about 20 m, height is about 1 m and crest length is about 800 m. They divide and interfinger with one another. In true plan view (see the diagram) the crests are aligned almost normal to the edge of the trains.

Continental shelf, English Channel.
Water depth about 59 m.

50°25′N 00°44′W
Sonograph W.E. × 3
Diagram W.E. × 1

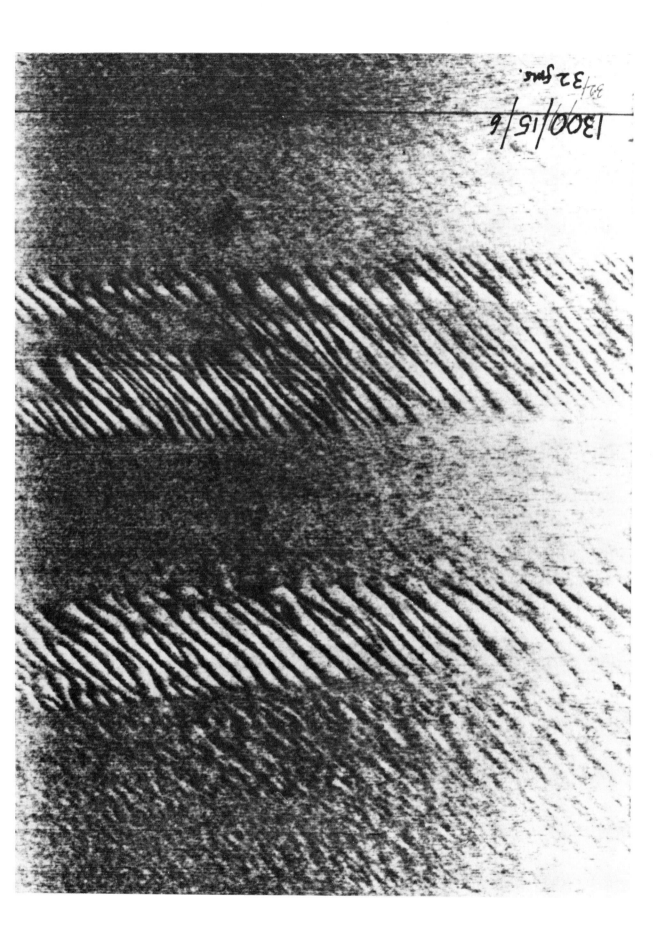

55

Gravel or Sand Waves

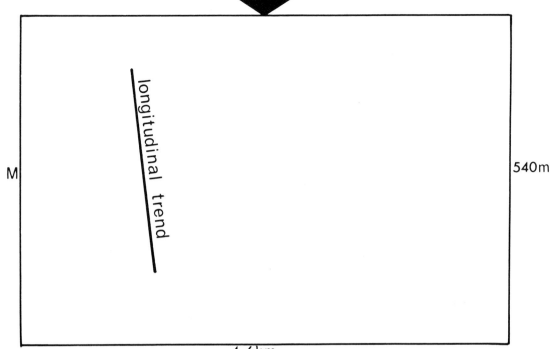

Fig. 45. An almost flat floor with irregular patches of gravel, and/or sand waves with a wavelength of about 10 m. The crests are mostly sinuous or herringbone in plan and extend approximately at right angles to the longitudinal trend of the train of waves.

Continental shelf, English Channel.
Water depth about 40 m.

50°21′N 01°17′W
W.E. × 5

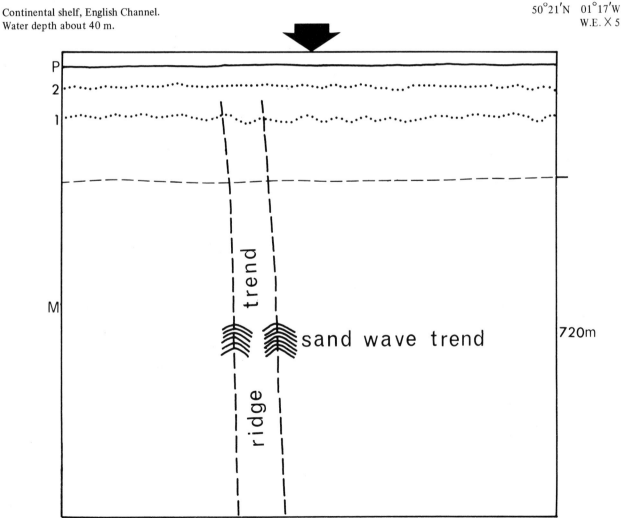

Fig. 46. Sand or gravel waves with a wavelength of about 10 m. The crests bend in unison over longitudinal ridges and troughs (seen best in *1*) producing a herringbone-like pattern. The longitudinal relief parallels the strongest currents. Multiple side lobes are prominent in this example.

Continental shelf, English Channel.
Water depth about 66 m.

50°06′N 01°27′W
W.E. × 4.5

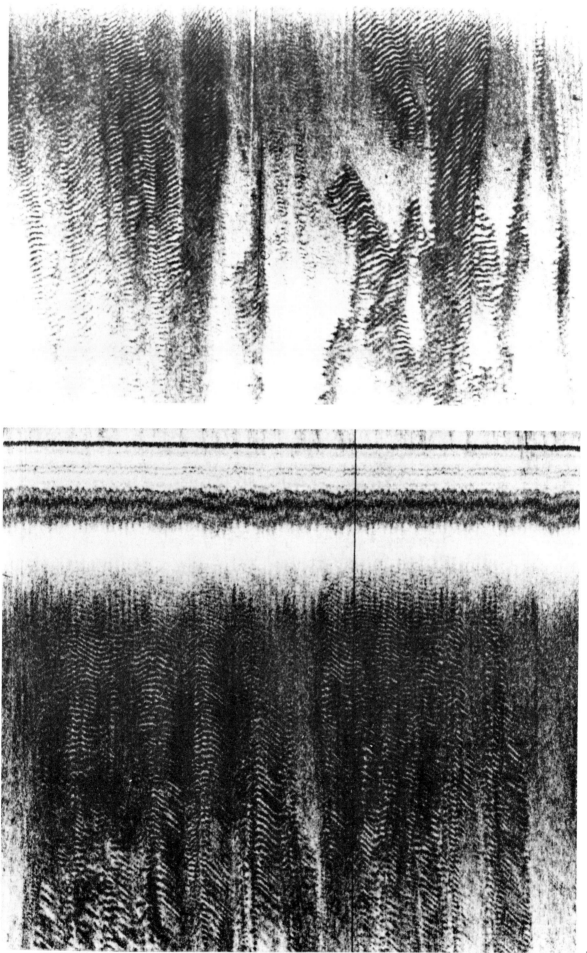

57

Sand Waves

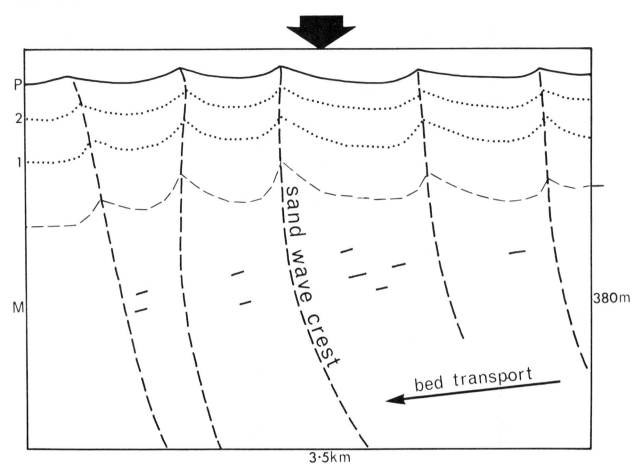

Fig. 47. Well-formed sand waves with a wavelength of about 700 m, height of about 18 m and an asymmetrical cross sectional profile at P, 2 and 1, the relatively steep slopes facing to the left. They are notable for being some of the tallest crests found around the British Isles and for the continuous curves between crests, suggesting an abundance of sand. There are faint suggestions of a secondary pattern extending normal to the crests, which can be seen by viewing the sonograph obliquely from left or right.

Continental shelf, Irish Sea.   53°22′N 05°35′W
Water depth in troughs about 75 m.   W.E. × 3.5

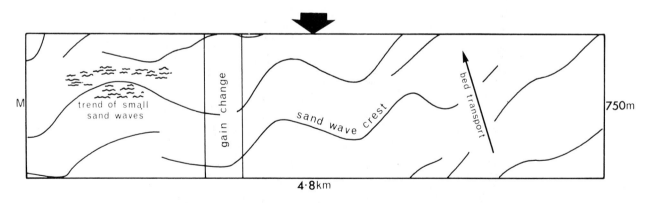

Fig. 48. Sinuous sand waves with a wavelength of about 300 m and height of about 5 m, in a region where the whole floor consists of sand. The asymmetrical cross-sectional profile of these sand waves is shown by the relatively narrow strip of the steep slopes (black) facing the top of the page compared to the wider band of the gentler slopes facing the bottom of the page. Much smaller sand waves with a wavelength of about 10 m are faintly shown. The general trends of the two sand wave sets are almost parallel. The diagram is a true plan representation.

Continental shelf, North Sea.   52°46′N 03°42′E
Water depth in troughs about 29 m.   Sonograph W.E. × 2.5
                                                              Diagram W.E. × 1

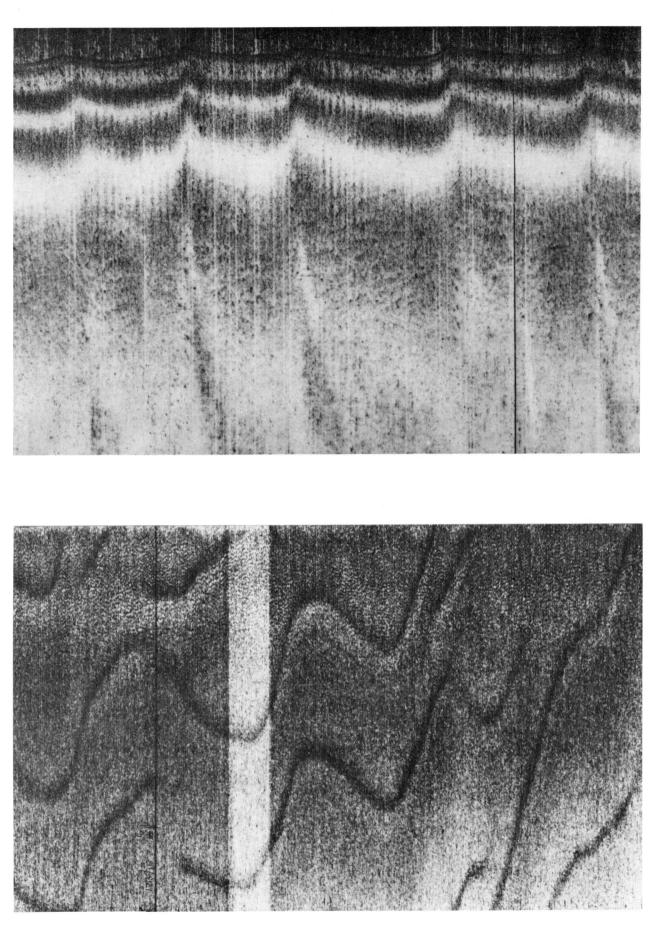

Sand Waves

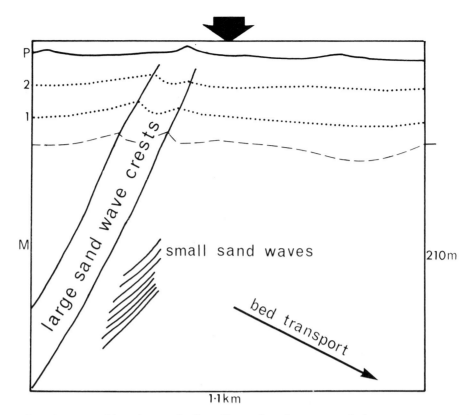

Fig. 49. Large and small sand waves masking a hummocky floor. The small sand waves meet the large ones at a true angle of about 20°, although this is accentuated on the sonograph by the width exaggeration.

Continental shelf, southern North Sea.   51°59′N   02°39′E
Maximum water depth 37 m.   W.E. × 3

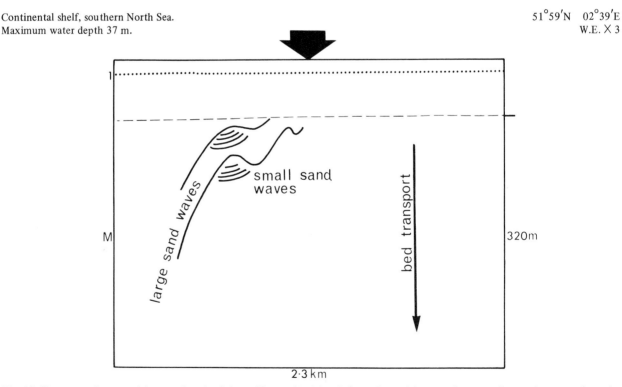

Fig. 50. Sinuous sand waves with a wavelength of about 70 m and height of about 4 m, with steep slopes tending to face away from the ship and casting strong shadows (white). The crests are separated from one another by a flat floor on which smaller sinuous sand waves (wavelength about 5 m and height less than 1 m) are present. The latter are locally arranged in concentric patterns in relation to the large crests, a pattern resembling ripple fans.

Continental shelf, Bristol Channel.   51°11.5′N   04°31′W
Water depth about 38 m.   W.E. × 4

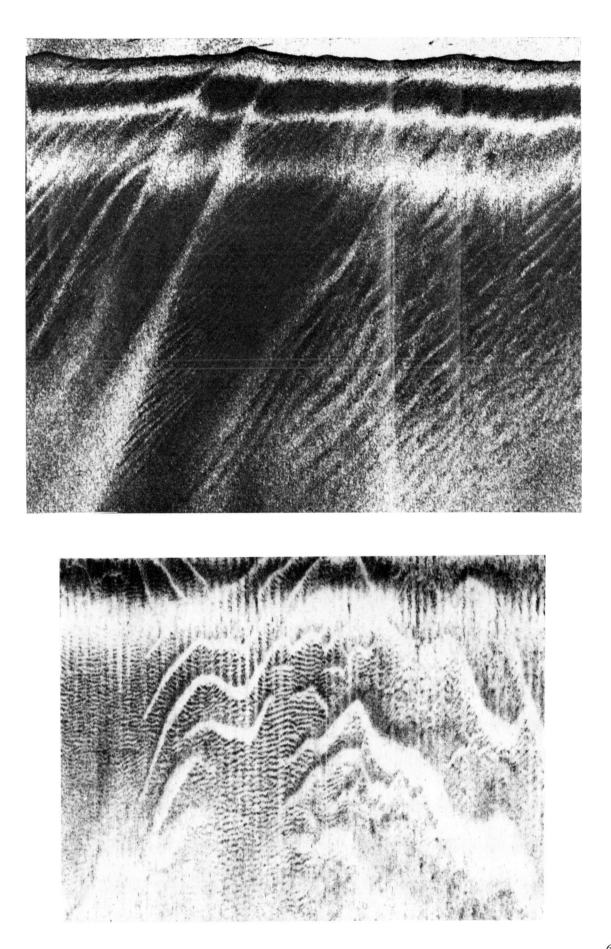

Sand Waves

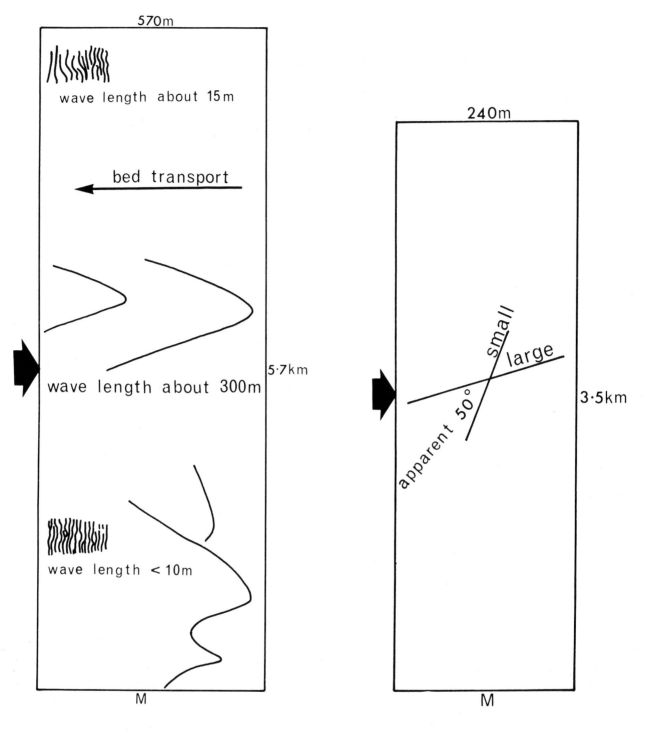

Fig. 51. Irregular to sinuous sand waves with a wavelength of about 300 m, carrying smaller sand waves with a wavelength of less than 10 m. On the flat floor beyond the large sand waves there are two isolated trains of small sand waves with a wavelength of about 15 m. The steeper slopes face the ship and appear black.

Continental shelf, North Sea.   52°53′N 03°20′E
Minimum water depth about 26 m.   W.E. × 3.5

Fig. 52. Sand waves, almost symmetrical in cross-sectional profile, with a height of up to 9 m and wavelength of about 400 m. These large crests are notable for the smaller sand waves (wavelength about 15 m and height less than 1 m) which trend obliquely over them from one side to the other. The difference in the trends of the two size groups probably indicates that the sand transport direction varies in keeping with the complex tidal flow around a nearby major headland. The true angular relationship is about 30°, after allowing for width exaggeration.

Continental shelf, Celtic Sea.   51°32.5′N 05°27.5′W
Minimum depth 55 m.   W.E. × 4.5

Sand Waves

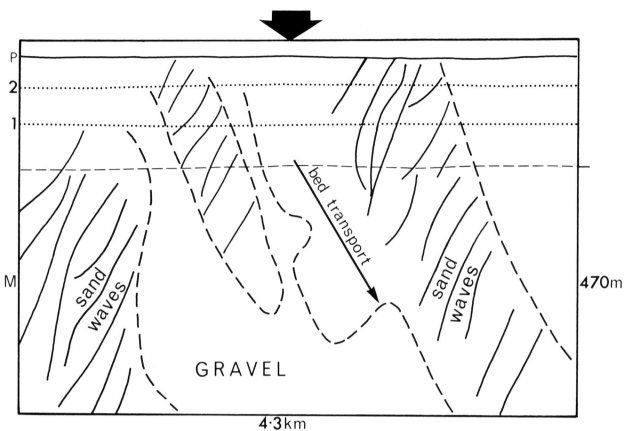

Fig. 53. Rather irregular trains of low sand waves with a wavelength of about 80 m, separated by a gravel floor. The wavy edges of *M, 1,* and *2* are due to a mechanical fault in the beam stabilisation system.

Continental shelf, Celtic Sea.
Water depth about 68 m.

50°49′N 05°25.5′W
W.E. X 4

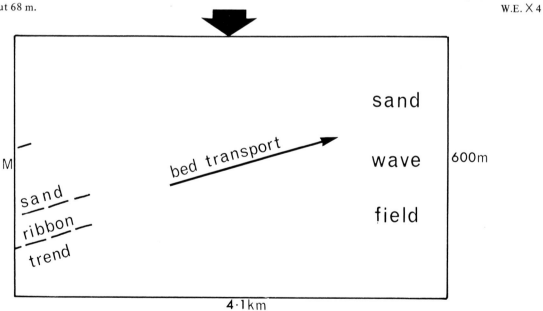

Fig. 54. Anastomosing, barchan-like sand waves occurring between sand ribbons (left) and a field of large sand waves (right) which are not clearly visible on the sonograph because of their trend perpendicular to the ship's course, but are shown by echo-sounder to be on average about 6 m high.

Continental shelf, Irish Sea.
Water depth about 71 m.

53°30′N 05°16′W
W.E. X 4

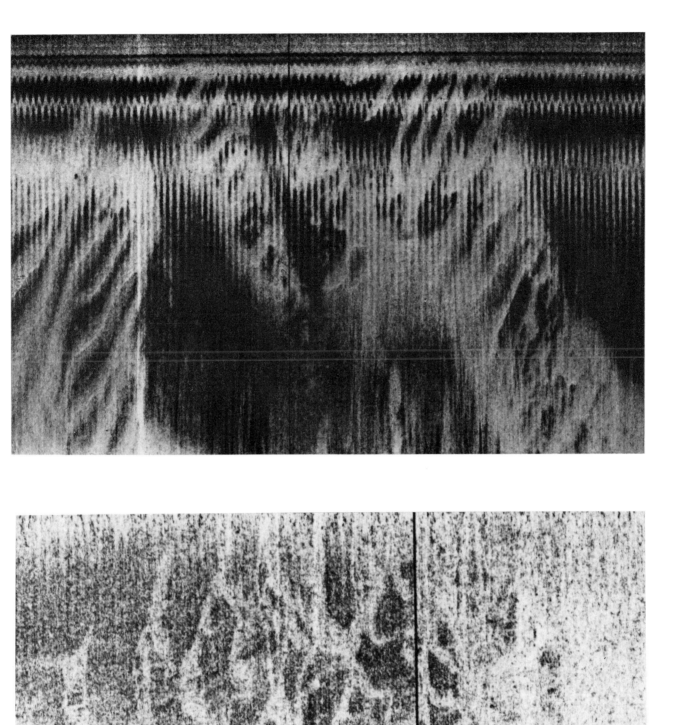

Sand Waves

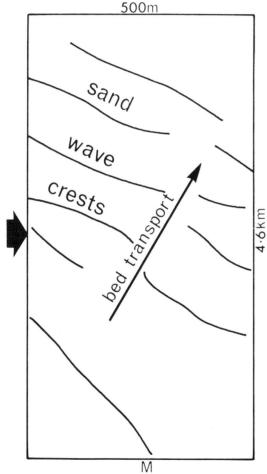

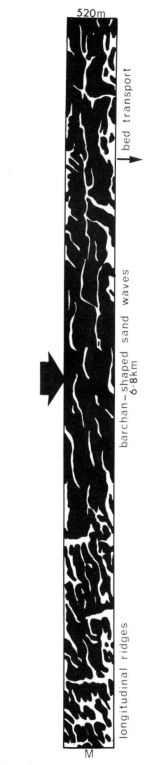

Fig. 55. Sand waves with asymmetrical profile (seen by echosounding), the steeper slope facing away from the ship showing as a white line, the more gentle slope, facing the ship, showing grey. The black patches within the troughs represent coarser ground swept clear of sand. Thus, the three tones on the sonograph are produced by a combination of relief and bottom texture.

Continental shelf, Celtic Sea.   48°49'N 09°11'W
Water depth about 157 m.   W.E. ✕ 2.5

Fig. 56. Barchan-like sand waves (pale tone) up to 2 m high, the wing tips of which tend to lead on to the middle of those down current from them. Some of the wings have been drawn out and united to form longitudinal ridges, perhaps the tidal equivalent of seif dunes of deserts.

Continental shelf, west of France.   47°26'N 05°14'W
Water depth about 135 m.   Sonograph W.E. ✕ 2.5
                                           Diagram W.E. ✕ 1

66

67

Sand Waves

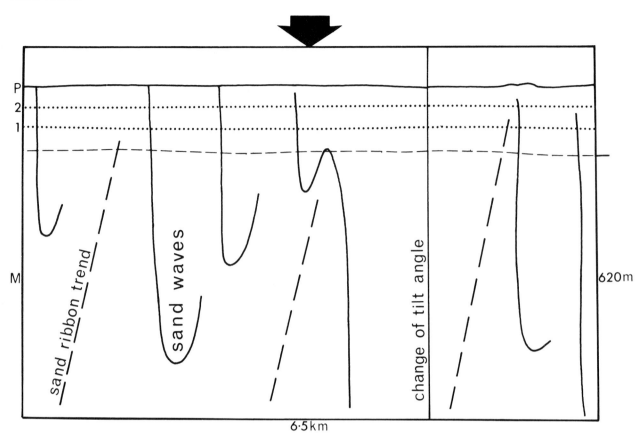

Fig. 57. Narrow, oblique sand waves (up to 7 m high and greater than 1 km long) perhaps analogous to the hooked dunes of the Arabian or other deserts. The sand waves and the just perceptible sand ribbons are separated by flat gravel floor.

Continental shelf, English Channel.   49°25′N   04°30′W
Water depth about 95 m.   W.E. × 4.5

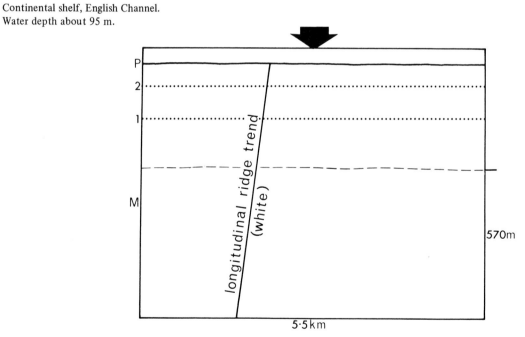

Fig. 58. Longitudinal ridges of sand greater than 1 km long, about 100 m wide and up to 4 m high, with separations of about 400 m, on a gravel floor. They are interpreted as the tidal equivalent of the seif dunes of a desert. The tidal ellipse is broad, implying some sideways as well as longitudinal sand movement. The scatter of small black blobs is due to fish shoals.

Continental shelf, Celtic Sea.   48°07′N   06°27′W
Water depth about 146 m.   W.E. × 4

68

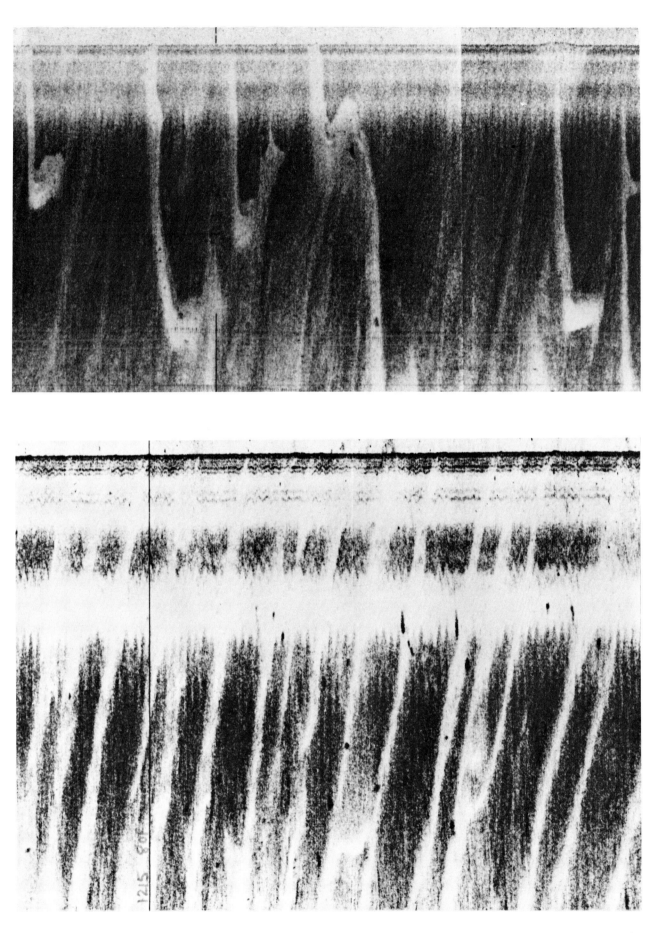

Sediment Patches

Transverse sand patches (light tone) up to 3 m thick, separated by gravel floors (dark tone). They have distinctive crescentic to ragged outlines, thought to indicate some mobility. The diagrams, in true plan view, show that these patches are rather similar in shape to those in Fig. 62–64, but viewed from a direction almost at right angles.

Continental shelf, Celtic Sea.

Fig. 59. Sand patches.
Water depth about 77 m.

51°29'N   06°56'W
Sonograph W.E. × 3.5
Diagram W.E. × 1

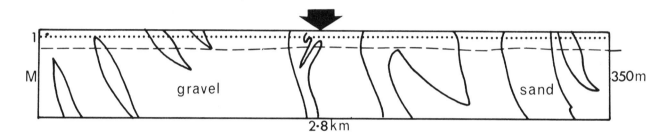

Fig. 60. Sand patches.
Water depth about 73 m.

51°35'N   07°16'W
Sonograph W.E. × 4
Diagram W.E. × 1

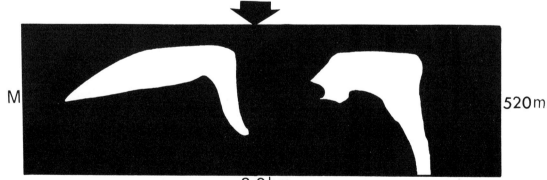

Fig. 61. Sand patches.
Water depth about 73 m.

51°41'N   07°14'W
Sonograph W.E. × 3
Diagram W.E. × 1

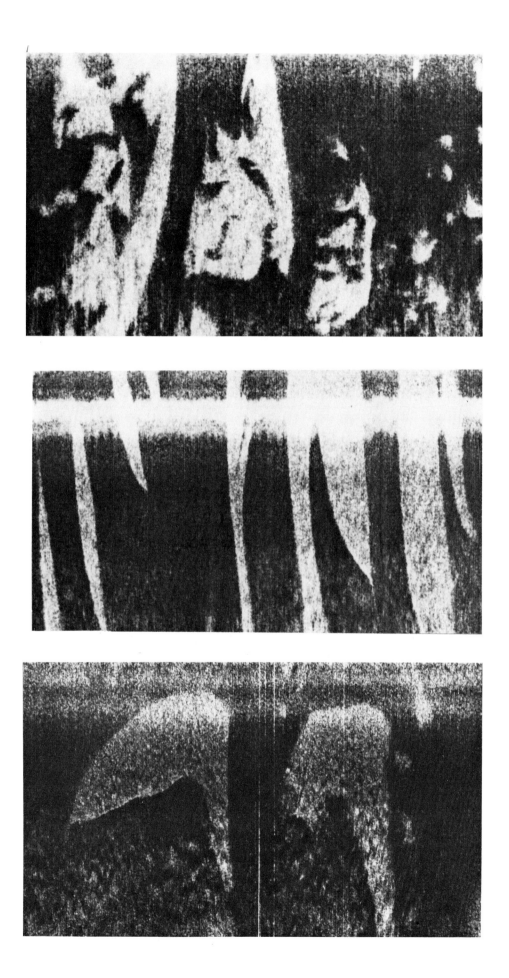

Sediment Patches

Transverse sand patches (light tone) up to 3 m thick, which may be subject to some movement. They are separated by gravel. Their distinctive shapes are, in fact, rather similar to those in Fig. 59–61 from the same region as will be seen by a comparison of the diagrams drawn in true plan view (except Fig. 64).

Continental shelf, Celtic Sea.

Fig. 62. Sand patches, The small-scale zig-zag pattern in the bottom half is a refraction effect. Water depth about 81 m.

51°24'N   07°25'W
Sonograph W.E. ✕ 2.5
Diagram W.E. ✕ 1

Fig. 63. Sand patches.
Water depth about 135 m.

!49°56'N   09°55'W
Sonograph W.E. ✕ 2.5
Diagram W.E. ✕ 1

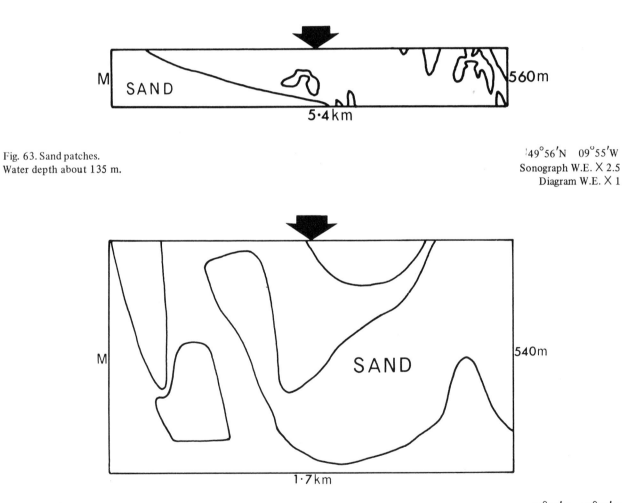

Fig. 64. Sand patches.
Water depth about 112 m.

50°00'N   09°29'W
W.E. ✕ 1.5

72

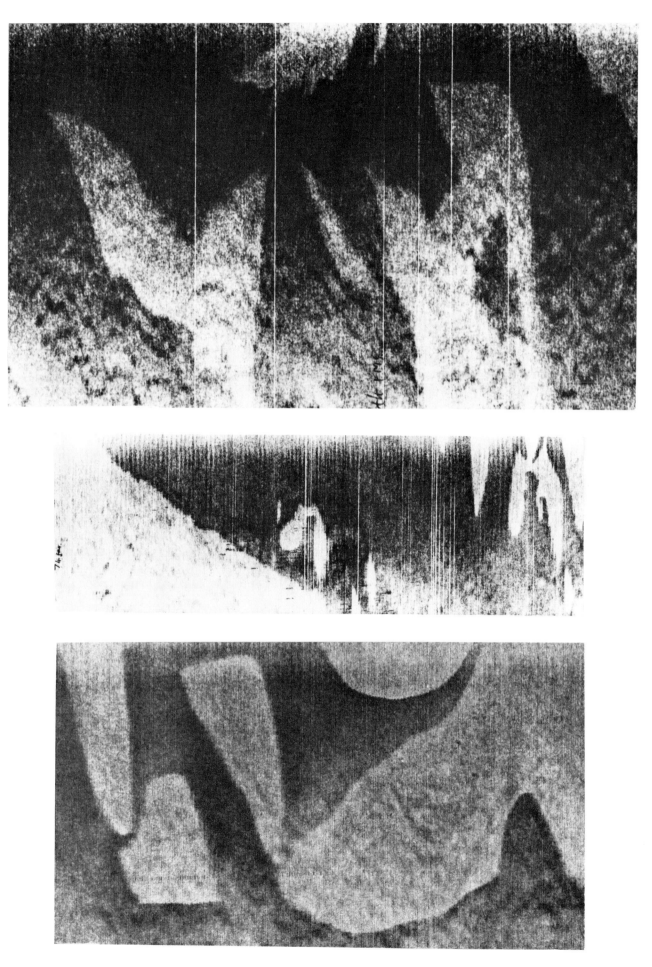

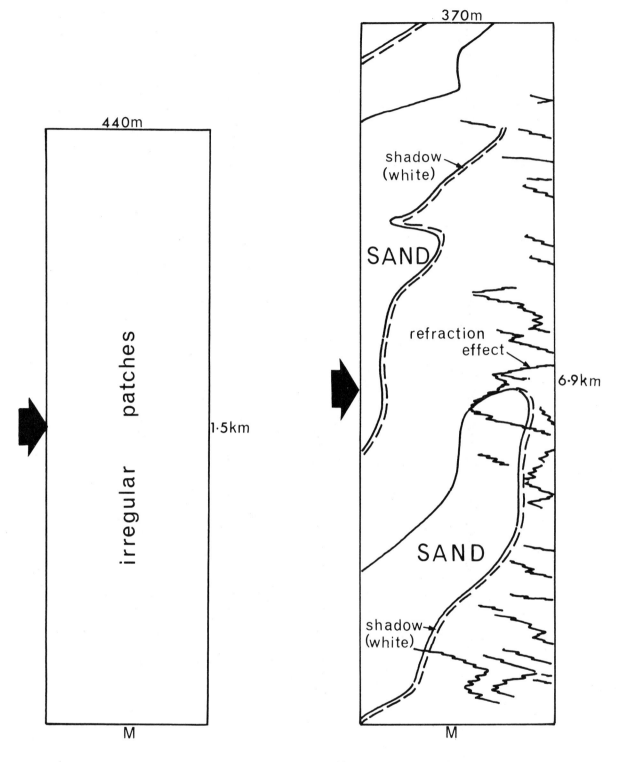

Fig. 65. Patches of sand and gravel having no markedly preferred orientation. Some of the small dark toned patches within the sandy areas (white) might possibly be coral.

Continental shelf, Persian Gulf.
W.E. about × 1
Water depth about 100 m.
Transit Sonar, by courtesy of Kelvin Hughes (Smiths Industries).

Fig. 66. Large, smooth-edged sand patches (light tone) up to about 2 m high, on a flat floor of coarser material (dark tone).

Continental shelf, west of Portugal.   40°35′N   09°05.5′W
Water depth about 80 m,   W.E. × 5

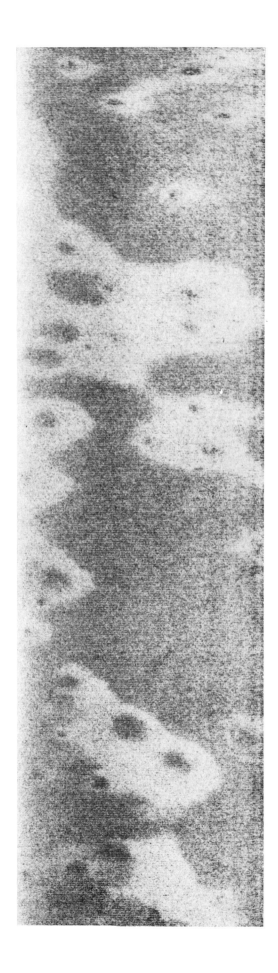

Sediment Patches

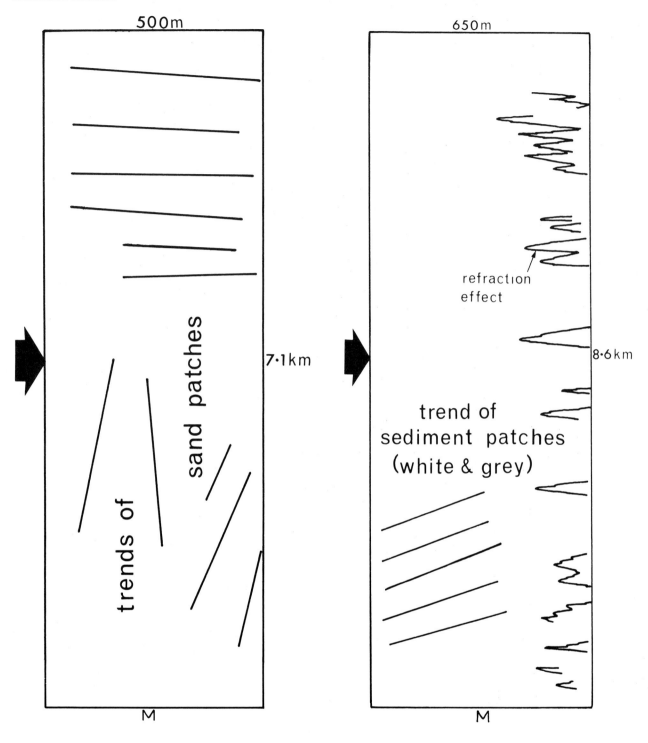

Fig. 67. Sand patches (with two trends) up to 3 km long, 2 m high and with a mean width of about 80 m separated by coarser floor (dark tone). They are remarkable for being elongated across a broad channel (bottom) and so are transverse to the main flow, whereas towards a large bank (top and off the sonograph) forming one edge of the channel they are parallel with the main current flow. The width exaggeration of the sonograph accounts for the apparent difference in width between the patches of the two trends.

Continental shelf, Celtic Sea.　　　49°14′N　10°W
Water depth about 130–150 m.　　　W.E. × 2.5

Fig. 68. A flat floor consisting of three materials of different textures. The fine-grained (pale tone) and coarser-grained (medium tone) longitudinal patches of sediment extend parallel with the main tidal flow, while the irregular shaped patches of darkest tone represent the coarsest material.

Continental shelf, Celtic Sea.　　　51°11′N　07°09′W
Water depth about 95 m.　　　W.E. × 4.5

Sediment Patches

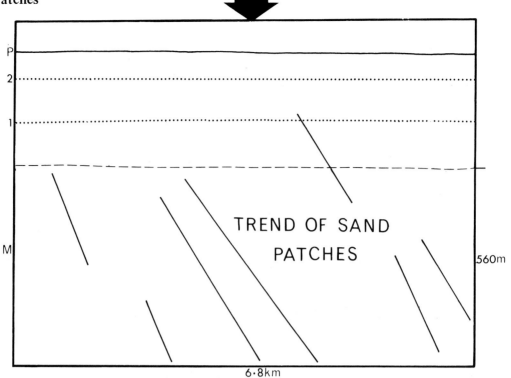

Fig. 69. Ragged-edged sand patches (light tone) with relief of about 2 m, apparently elongated parallel with the main tidal flow.

Continental shelf, west of Ireland.
Water depth about 135 m.

52°30′N   11°28′W
W.E. × 4.5

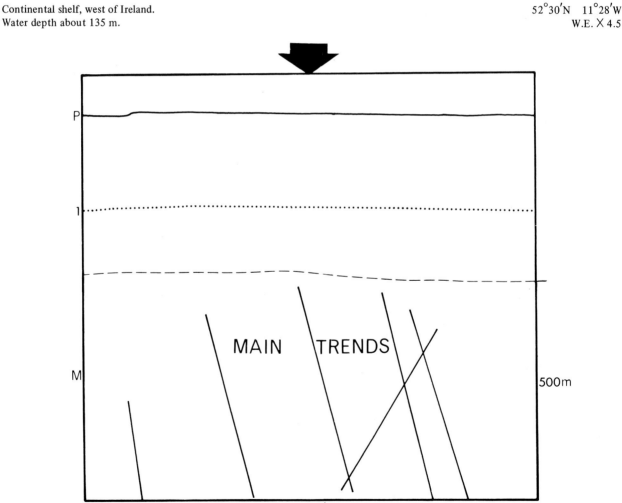

Fig. 70. A gravel floor (dark tone) almost buried by sand, but with enough patches showing through to indicate that the sand has a linear trend and reaches up to about 2 m above the gravel. These sand patches are thicker and more complete than nearby ones of similar trend shown in Fig. 69.

Continental shelf, west of Ireland.
Water depth about 150 m.

52°24′N   11°29′W
W.E. × 4.5

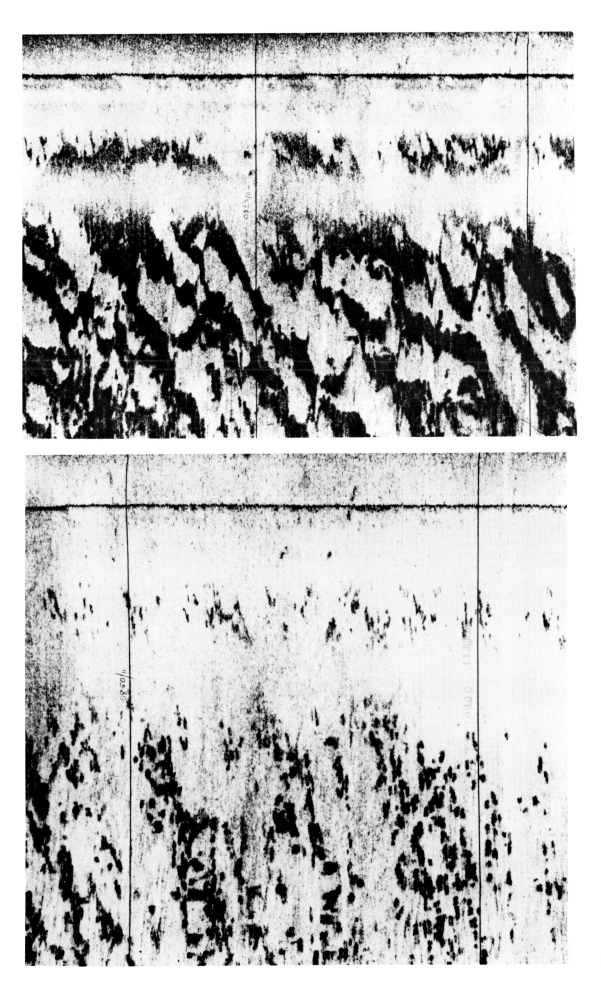

Sediment Patches

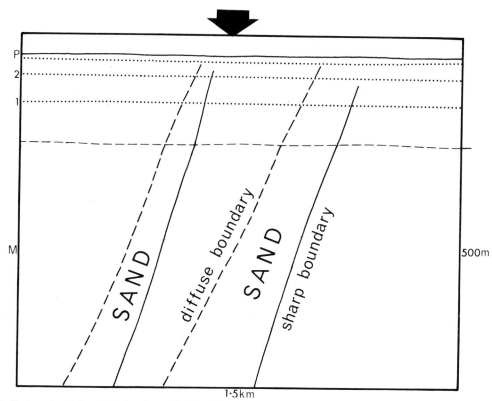

Fig. 71. Longitudinal sand patches (light tone) notable for their asymmetry in plan view. The fairly sharp boundary on one side and a gradational one on the other, probably indicates progressive decrease in sand thickness across the patch from less than about two metres to zero. The small black blobs may be rock outcrops, groups of boulders or fish.

Continental shelf, Celtic Sea.
Water depth about 53 m.

50°38.5'N   05°14'W
W.E. ✕1.5

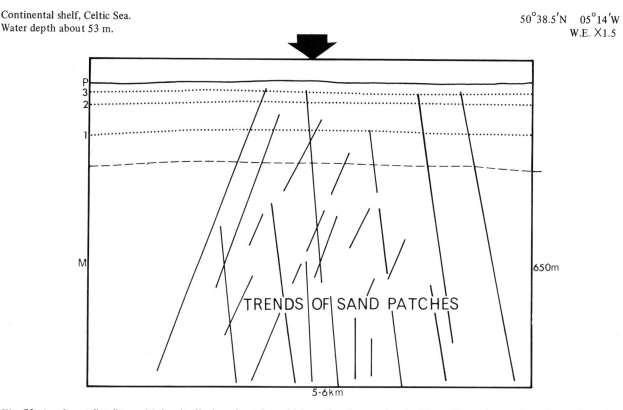

Fig. 72. An almost flat floor with longitudinal sand patches which are locally associated with smaller and more irregular sand patches oriented almost normal to them (after making allowance for width exaggeration). The apparent curvature of these patches near to *P* indicates rather nicely that the non-linearity of the width scale is only a nuisance near *2,1* and P.

Continental shelf, west of Ireland.
Water depth about 95 m.

54°34'N   10°04'W
W.E. ✕ 3.5

Sediment Patches

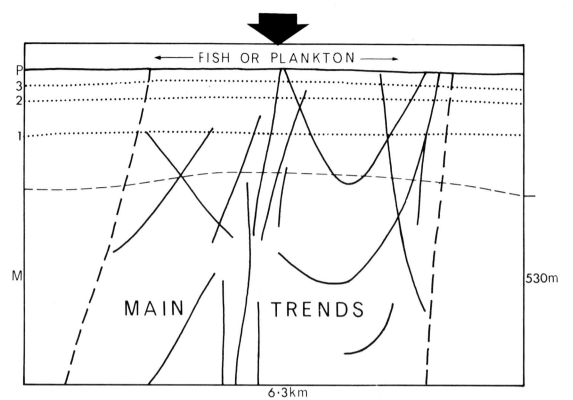

Fig. 73. Long narrow, straight and curving trends, revealed by the presence of sand, in an almost flat but texturally rougher floor. Gravel is also present but these trends are most suggestive of structures in rock, which have perhaps been complicated and emphasised by ice scour.

Continental shelf, west of Scotland.
Water depth about 150 m.

57°34′N   08°48′W
W.E. × 4

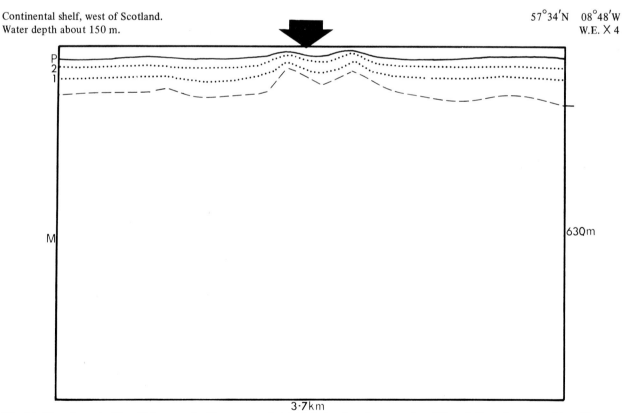

Fig. 74. Mounds up to 20 m high, capped with coarse material (dark tone) such as gravel, pebbles, shells and some boulders. They are surrounded by a smooth sand floor (light tone). The mounds are probably of glacial origin while the sand was probably spread out by the present sea.

Continental shelf, Firth of Clyde, Scotland.
Water depth in profile about 18 to 30 m.

55°11′N   04°59.5′W
W.E. × 3

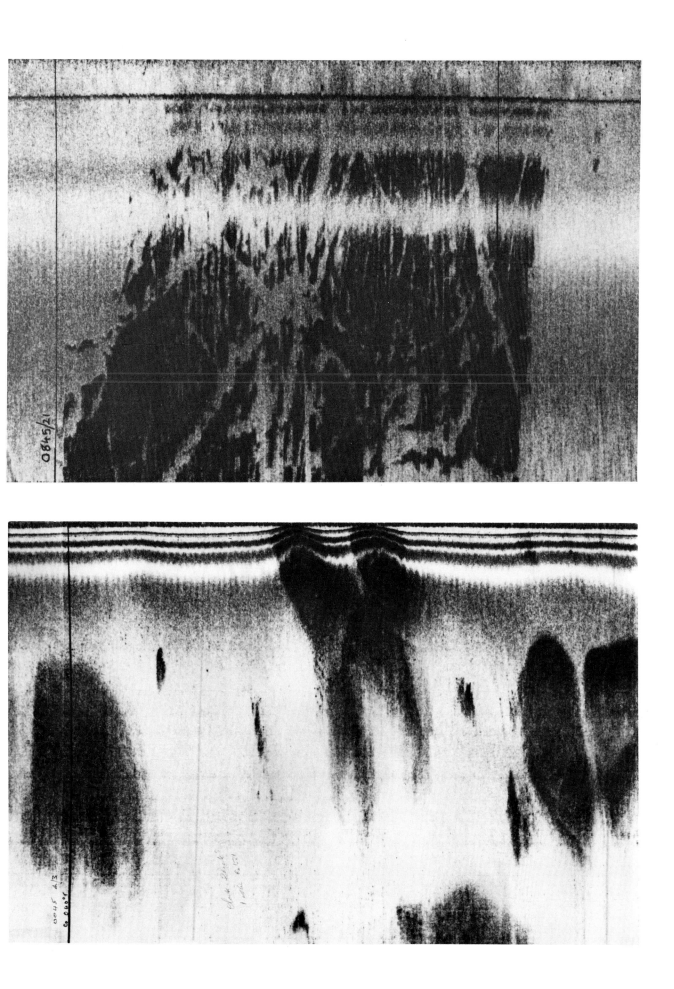

Sediment Boundaries

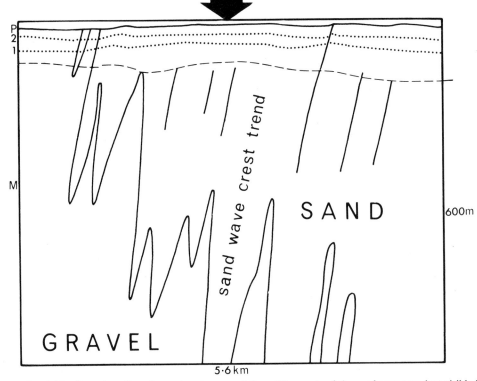

Fig. 75. Sand waves of variable dimensions fingering out on to a gravel floor. The crests of the sand waves are just visible in profile, with the steeper slopes facing the left hand side of the sonograph. In plan view the leading edges of the sand waves are more sharply defined than those of the trailing edges.

Continental shelf, English Channel.
Water depth about 24—35 m.

50°39.5′N   00°04′E
W.E. ✕ 6

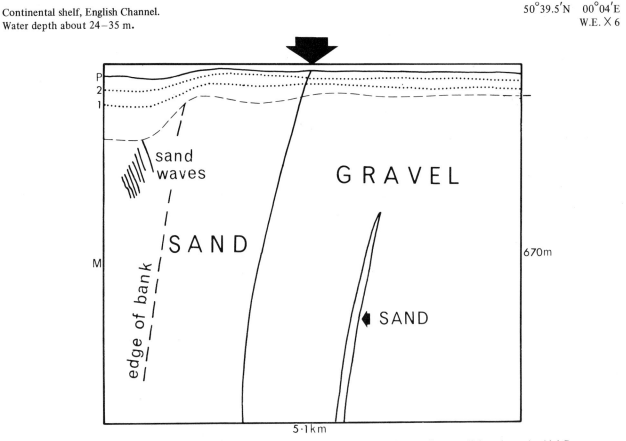

Fig. 76. A sharp boundary between sand and gravel occurring on the top of a bank and extending parallel to the main tidal flow.

Continental shelf, English Channel.
Water depth on bank about 20 m.
Water depth on surrounding floor about 38 m.

50°44′N   00°36′E
W.E. ✕ 6

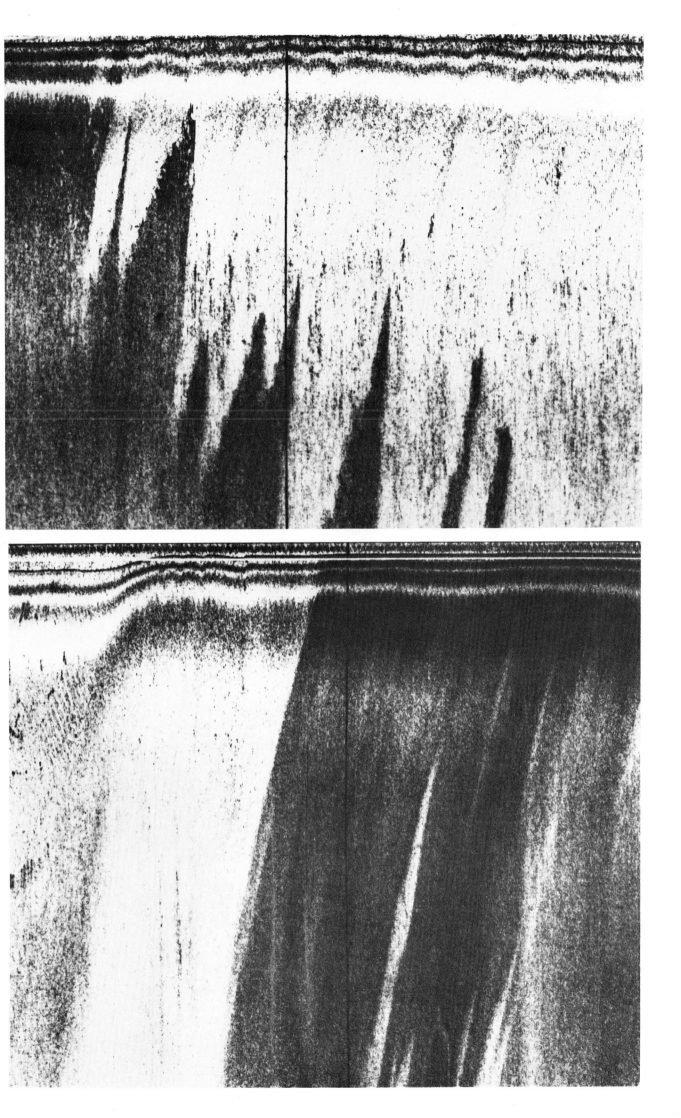

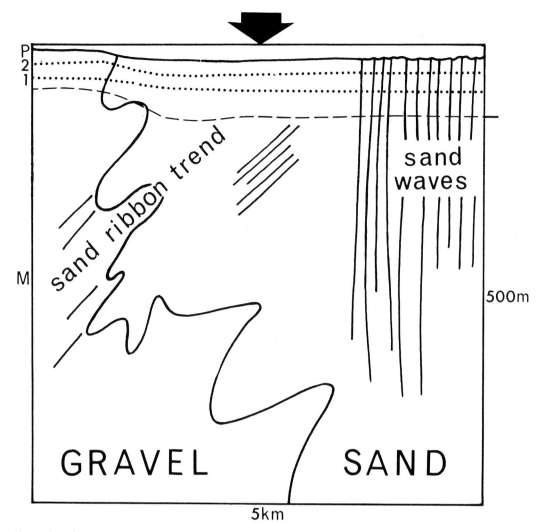

Fig. 77. The lobate edge of a thin sheet of sand flanking a "shingle" bank and extending transverse to the main tidal flow. The boundary interrupted by just perceptible sand ribbons, while gravel floor can be detected within the troughs of the sand waves.

Continental shelf, English Channel.　　　　　　　　　　　　　　　　　　　　　　50°44.5′N　00°37′E
Water depth on the bank about 17 m.　　　　　　　　　　　　　　　　　　　　　　W.E. X 7.5
Water depth on the surrounding floor about 26 m.

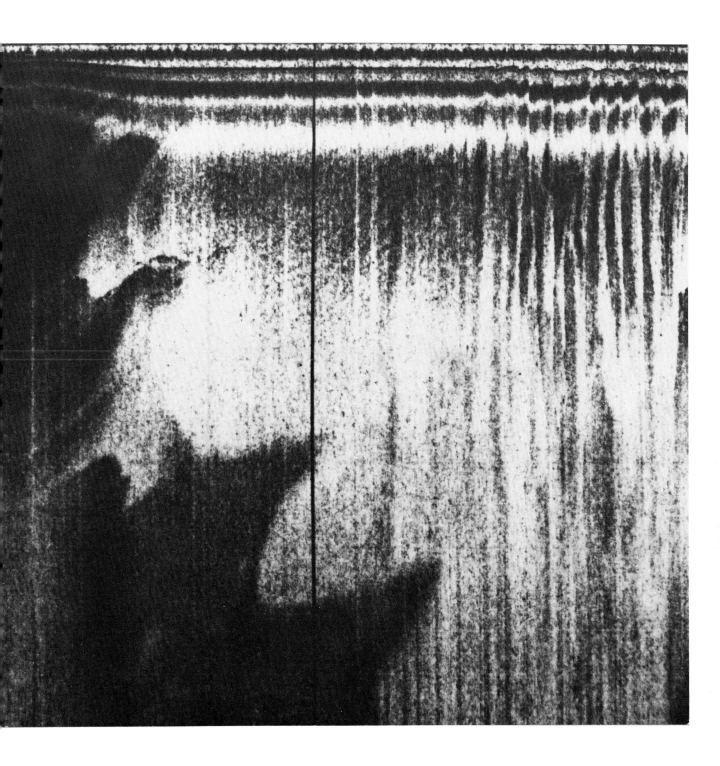

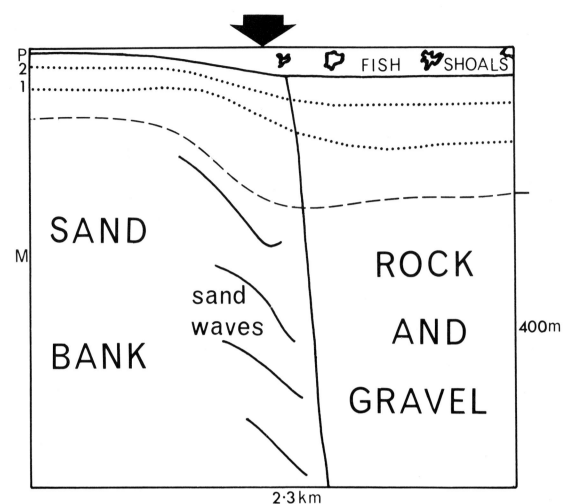

Fig. 78. A sharp boundary between a tall bank of sand and a floor of rock and gravel. The large sand waves on the side of the bank are oblique to its outer limit.

Continental shelf, Bristol Channel.
Water depth over bank about 23 m.
Water depth over deeper floor about 50 m.

51°10.5′N   04°46′W
W.E. × 3

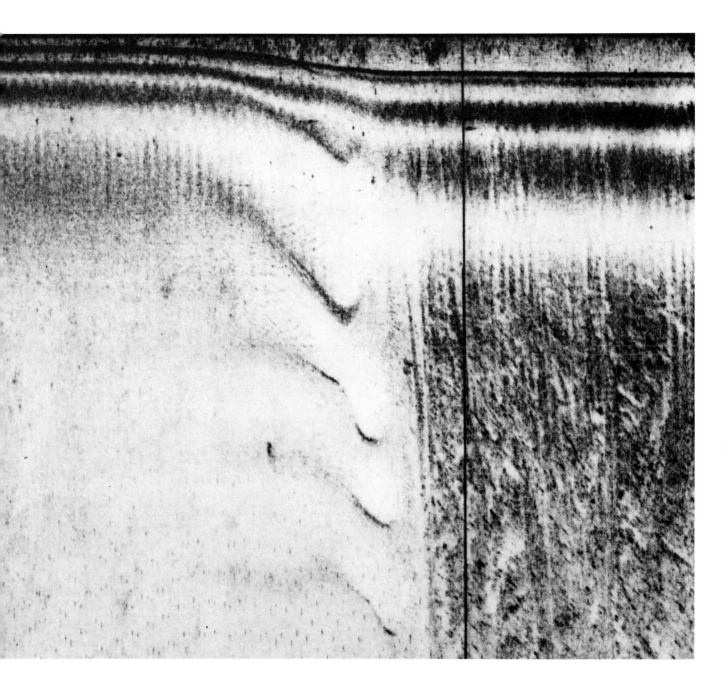

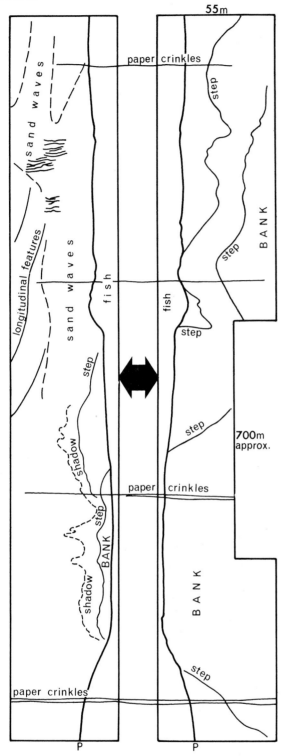

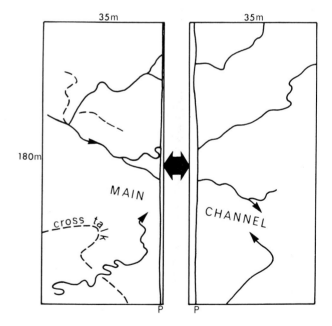

Fig. 79. A pair of sonographs, obtained by looking to both sides simultaneously, showing an estuarine bank and channel. The edge of the bank is notable for the low steps extending along it. On the deeper-lying floor there are small sand waves, together with longitudinal bed forms, possibly sand ribbons.

Thames Estuary, England.
Water depth 6.5–23 m.   W.E. about × 2
E.G. and G. Dual side-scan sonar. By courtesy of Hunting Surveys and Consultants Ltd.

Fig. 80. A pair of sonographs obtained by looking to both sides simultaneously, showing sinuous tidal channels on an estuarine tidal flat. One sonograph is marred by "cross talk" between the two transducers.

Tamar Estuary, Plymouth, England.   W.E. × 2
E.G. and G. Dual side-scan sonar. By courtesy of Hunting Surveys and Consultants Ltd.

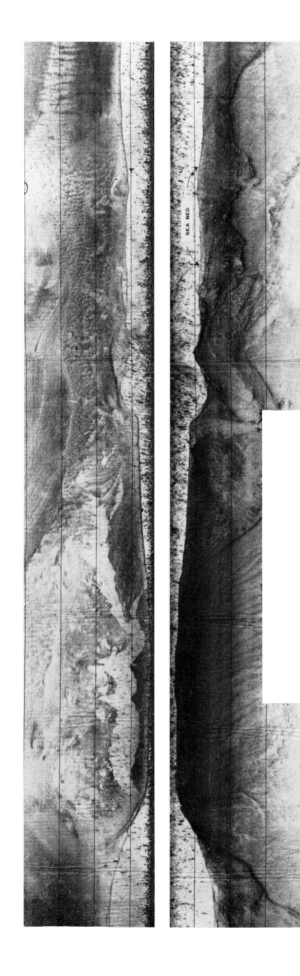

Coastal Features

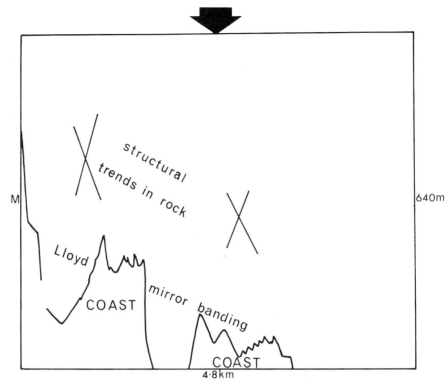

Fig. 81. A ragged coast whose outline seems to follow the same trends as the rock outcrops on the adjacent sea floor. Loose sediment mantles the rock on the floor of the bays and below the deeper water.

Continental shelf, Hong Kong.   22°20.8′N   114°0.7′E
Maximum water depth 42 m.   W.E. × 6
Towed Surveying Asdic. By courtesy of the Physics Dept., Hong Kong University.

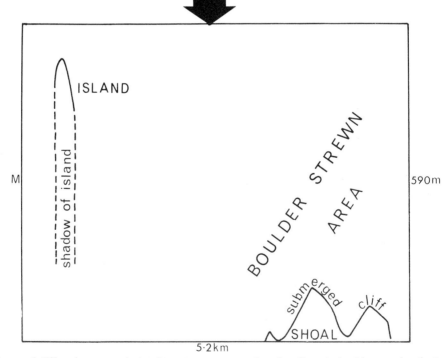

Fig. 82. An island, submerged cliffs and numerous isolated targets near a coast where 2-m diameter boulders are abundant.

Continental shelf, Hong Kong.   22°21′N   114°16.4′E
Maximum water depth about 18 m.   W.E. × 7
Towed Surveying Asdic. By courtesy of the Physics Dept., Hong Kong University.

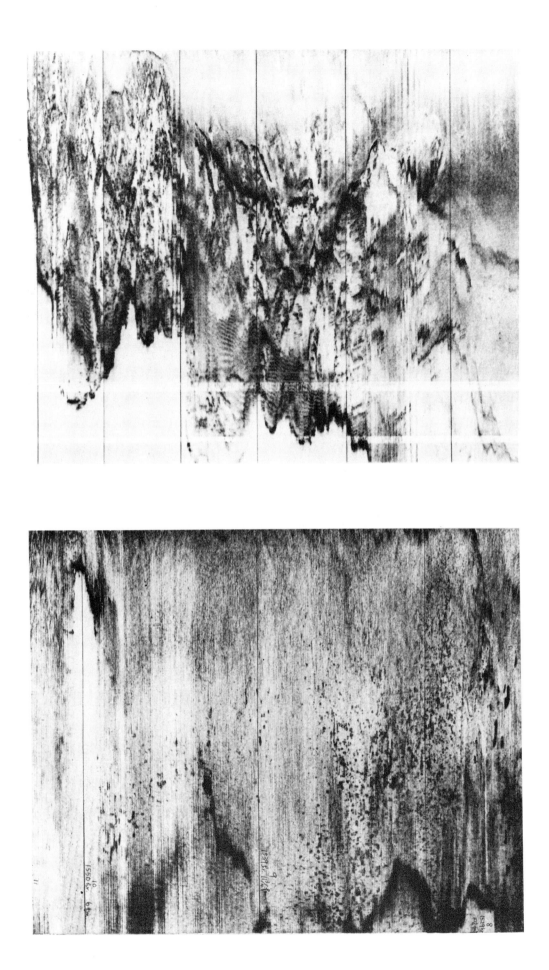

93

Ice Scour

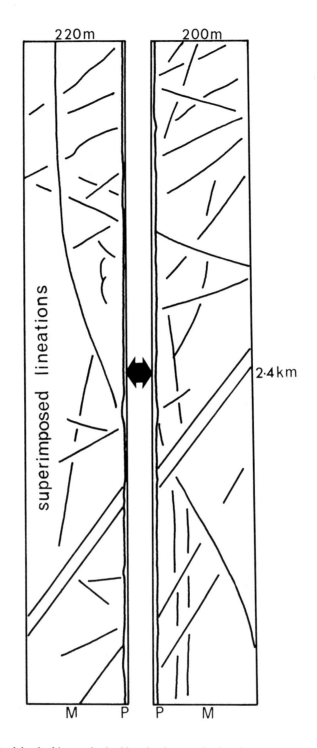

Fig. 83. A pair of sonographs obtained by looking to both sides simultaneously showing a criss-cross pattern of straight or curving furrows ploughed into muddy sediment by drifting sea ice. The furrows appear to have a more or less random orientation, and have been superimposed in a time sequence which can be worked out by inspection of intersections. They sometimes occur in parallel groups, are up to about 25 m across and 5 m deep, and are traceable for up to 2 km on this sonograph pair. The rims are often raised due to the shoving aside of sediment by the moving ice.

Continental shelf, about 50 km from Tuktoyaktuk Peninsula, Beaufort Sea, Canadian Arctic.
Water depth about 60 m.
E.G. and G. Dual side-scan sonar. By courtesy of Dr B.R. Pelletier, Bedford Institute of Oceanography, Canada.

W.E. × 1.5

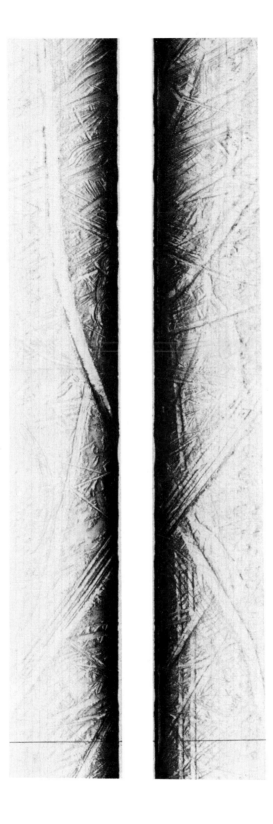

Coral

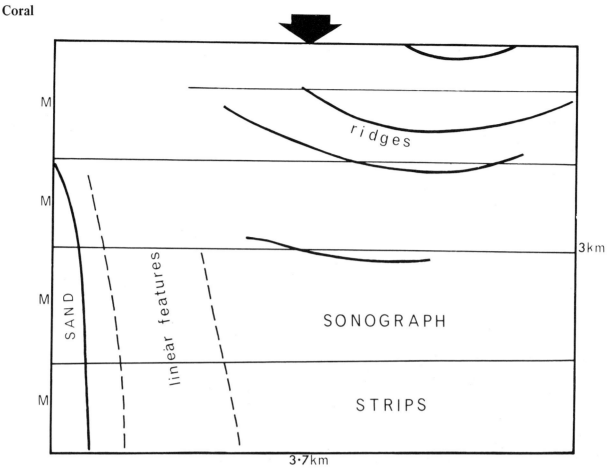

Fig. 84. A sonograph mosaic taken near the rim of a submerged atoll where colonies of coral (dark-toned irregular shapes) are separated by coral sand and coarser debris. The main curved features at the upper right of the figure are either coral or coral covered ridges.

Macclesfield Bank, South China Sea.   16°05.5′N   114°29′E
Water depth about 20–40 m.   W.E. X 1
Kelvin Hughes Towed Surveying Asdic. By courtesy of the Physics Department, Hong Kong University.

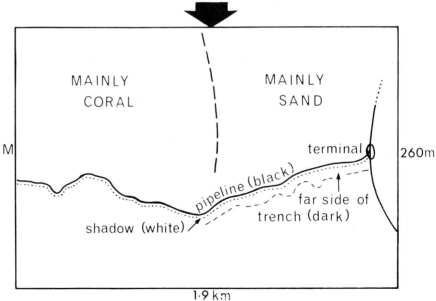

Fig. 85. Patchy sediments and small, irregular, strong reflections attributed to coral. Three pipelines converge on a terminal: the strong reflection behind the longest pipeline suggests that it is lying in a shallow trench.

Continental shelf, Persian Gulf.
Water depth about 20 m.   W.E. X 5
Towed Surveying Asdic. By courtesey of British Petroleum Co. Ltd.

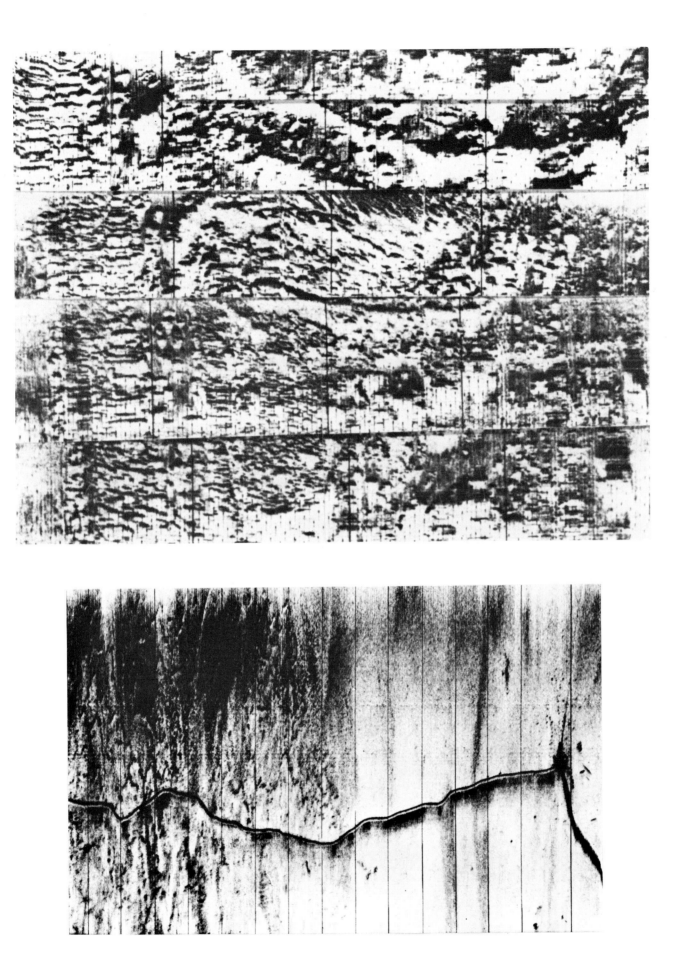

Coral

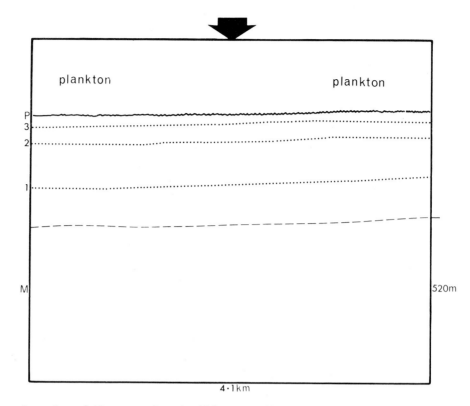

Fig. 86. An anastomosing pattern of ridges up to about 4 m high, separated by a sediment floor (light tone), in an area where abundant cold water corals *(Lophelia prolifera)* occur.

Western flank of Rockall Bank, northeast Atlantic.
Water depth about 235 m.

57°18′N 14°35′W
W.E. × 3

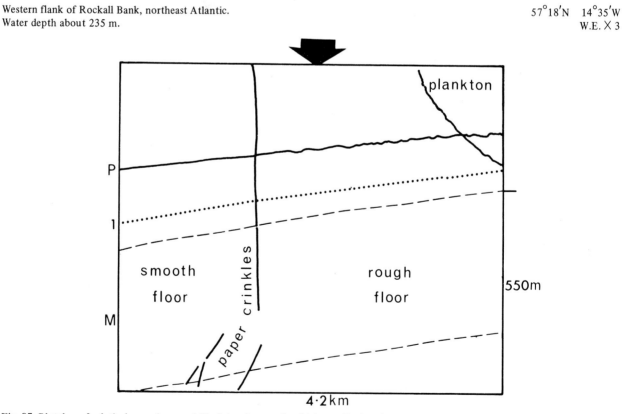

Fig. 87. Blotches of relatively rough ground (dark tone) up to 4 m high, ascribed to deep-living coral, surrounded by sand floor.

Continental slope, northwest of Morocco.
Minimum water depth 315 m.

35°35′N 06°31′W
W.E. × 3

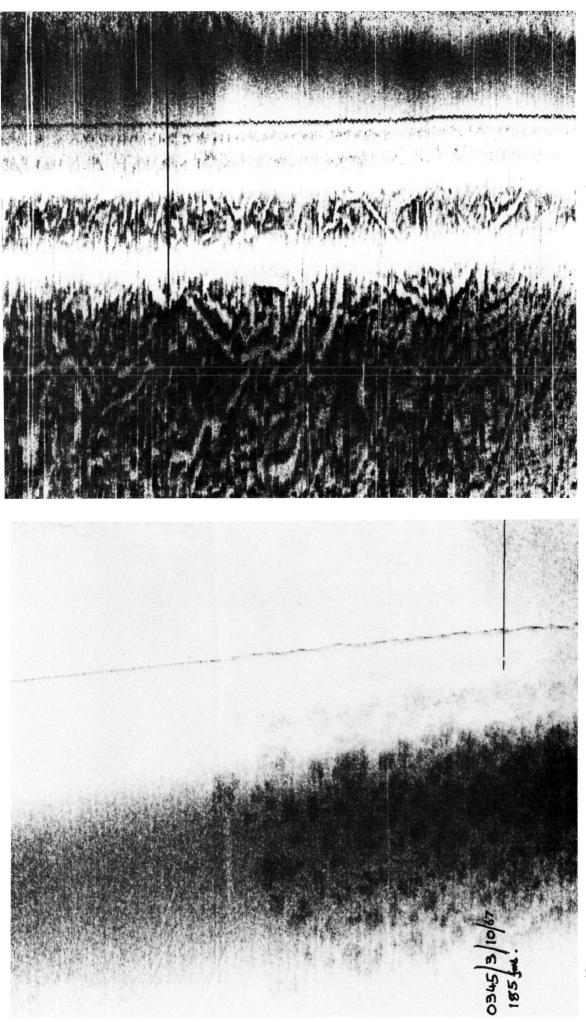

0345 / 3 / 10 / 67
185 fms.

99

UPPER CONTINENTAL SLOPE

Rock

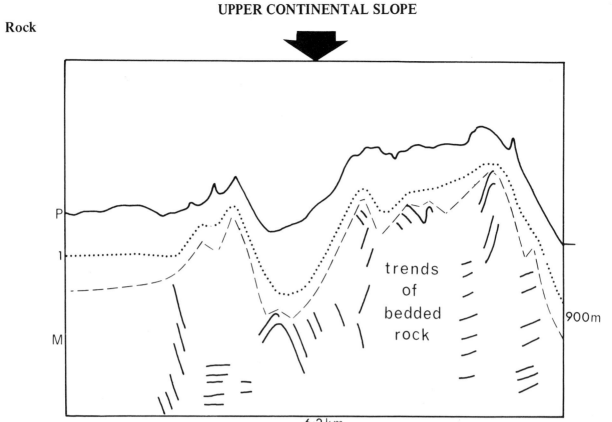

Fig. 88. A rugged portion of continental slope made of bedded rock, as indicated by the small parallel ridges with shadows beyond them. There is some sediment cover at the left hand side.

Continental slope, Strait of Gibraltar.   35°56′N 05°45′W
Minimum water depth 230 m.   W.E. X 3

Fig. 89. Irregular shaped patches of rough ground (dark tone), near to the top of the continental slope, that are attributed to rock outcrops, but are devoid of obvious signs of bedding. The intervening hollows are partly filled with loose sediment. Those dark areas shown by shading on the diagram represent rough floor lying beyond the maximum recorder range (Fig. 1), which is overprinted during the next sweep of the 'stylus'.

Continental slope, Gulf of Cadiz.   36°34′N 07°44′W
Maximum water depth about 350 m.   W.E. X 3.5

Sand Waves

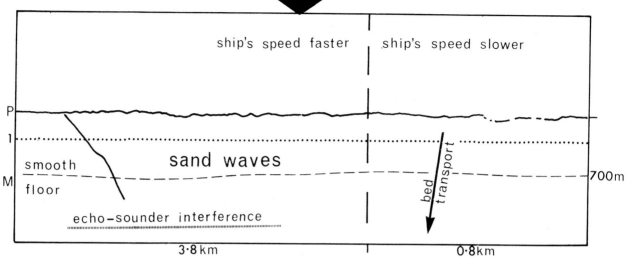

Fig. 90. A field of sand waves built by the relatively dense undercurrent flowing out of the Mediterranean Sea and then along the continental slope. The sand waves have short sinuous crests, shown in almost true plan view on the right of the sonograph. Their steeper slopes face away from the ship.

Continental slope, Gulf of Cadiz.   36°15'N   06°55'W
Water depth about 720 m.   W.E. Left × 2
                                                                   Right × 1

Cliffs

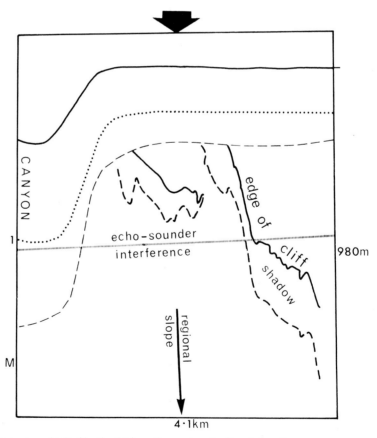

Fig. 91. Cliffs, whose upper edges are emphasised by the shadows they cast on the floor below them.

Continental slope off Barcelona, Spain.   41°23.5'N   02°40'E
Minimum water depth 170 m.   W.E. × 4.5

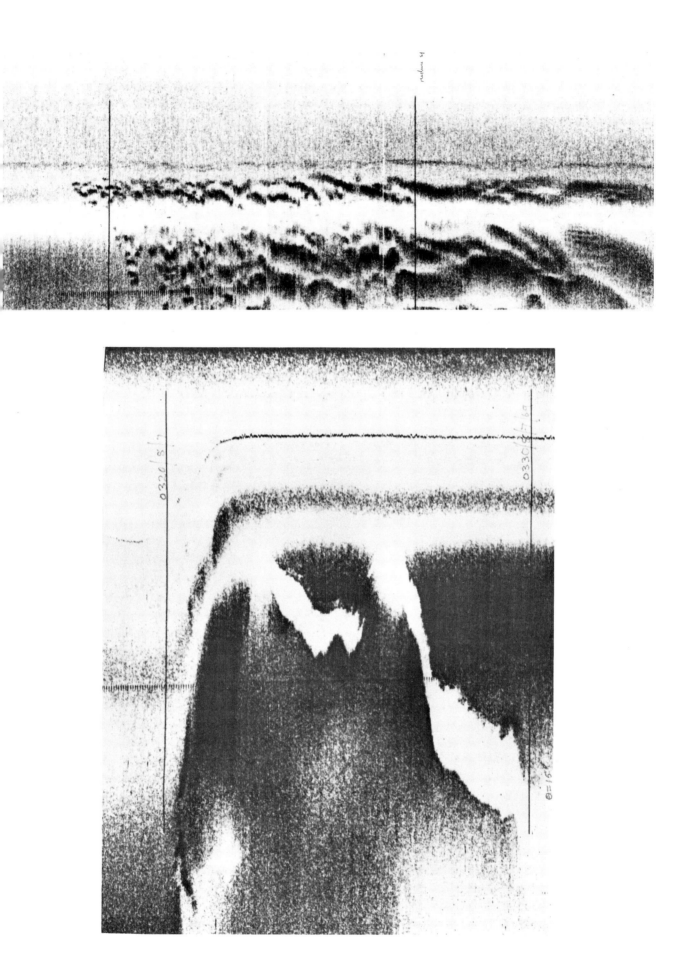

Slumps

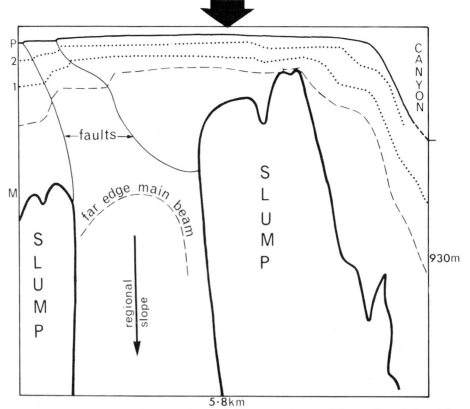

Fig. 92. An unusual pattern of alternate bands of rough and smooth floor near the top of the continental slope. The rough bands are attributed to ground which has been deformed into transverse ruckles during downslope movement as slumps, although they could possibly be rock. The smooth floor is crossed by two faint, curved lines, possibly traces of young faults associated with the slumping.

Continental slope off Algeria, Mediterranean.   36°55'N 03°41'E
Minimum water depth 88 m.   W.E. × 5

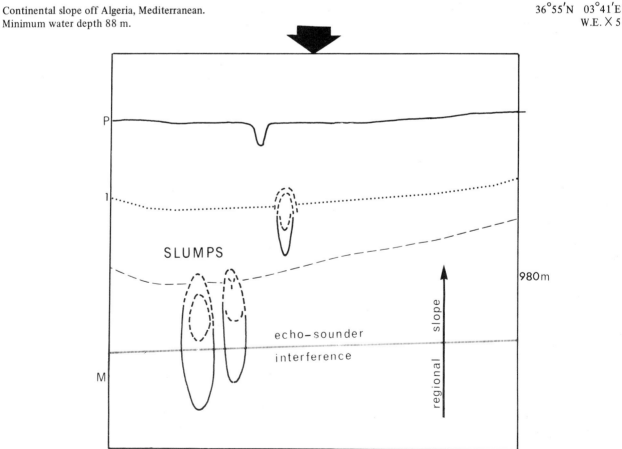

Fig. 93. A view looking up towards the top of a continental slope, whose smoothness is broken by small, almost circular relief features attributed to slumping. In profile, the slump scar is about 25 m deep, 250 m wide and has a curved lip. In plan view, the scars are relatively steep, rough, and face down the continental slope. The front of each slump has one or two ridges which cast shadows (white).

Continental slope, Mallorca, Mediterranean.   39°37'N 02°07.5'E
Water depth about 200 m.   W.E. × 4.5

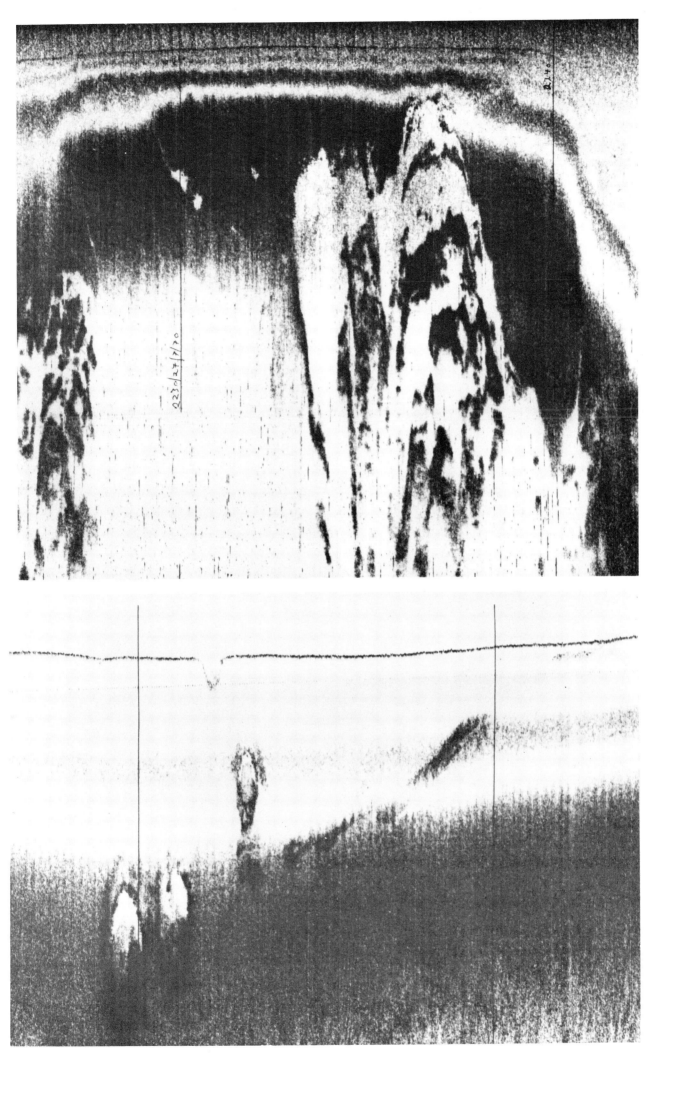

Submarine Canyons

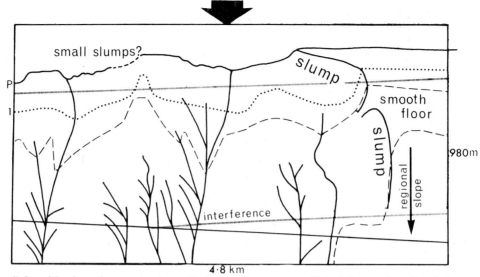

Fig. 94. Complex relief resulting from the concurrent growth of submarine canyons with both large and small scale slumps; alongside is a portion of undissected continental slope. The larger canyons reach up to the continental shelf while minor ones originate lower down. There are numerous gullies in the canyon walls. Small slumps also seem to be present in the canyon walls as shown in profile P, and may also occur on the flattish ground between the main canyons. A larger slump (or slumps) towards the right, lies below arcuate scars (in shadow).

Continental slope, Celtic Sea.
Least water depth on the continental shelf about 175 m.

48°46'N 10°16'W
W.E. X 2.5

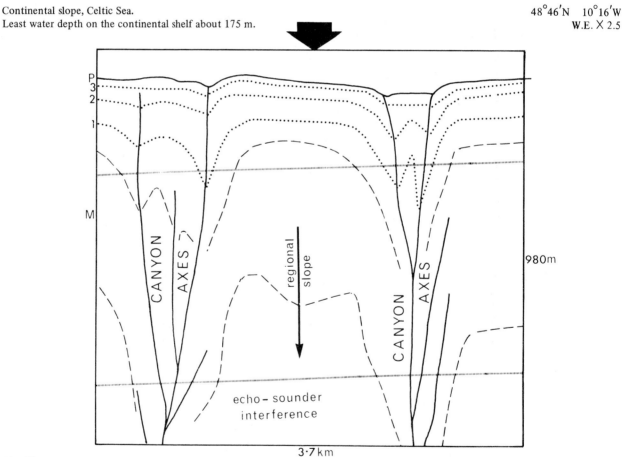

Fig. 95. Two canyons near to the edge of the continental shelf. The canyons are up to about 150 m deep in the profile P, but deepen rapidly away from the ship as shown by the progressively greater excursions of the more distant side lobes 3, 2 and 1 and by the edge of the main beam. There are some gullies in the canyon walls. Fig. 96 and 97 show stages in relative canyon development.

Continental slope off Algeria, Mediterranean.
Maximum water depth about 130 m.

36°10'N 00°13'E
W.E. X 3

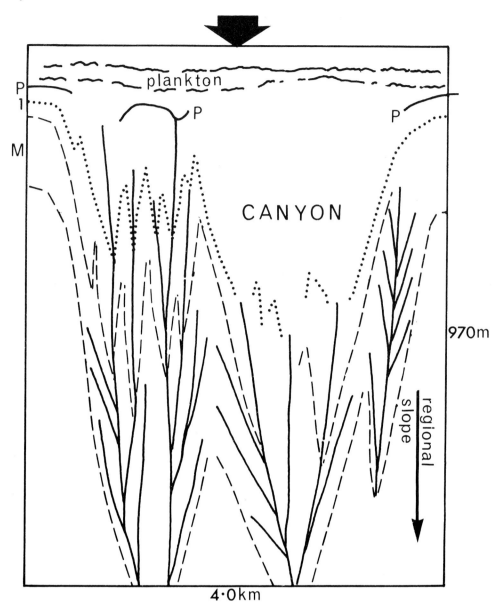

Fig. 96. Two canyons with small, almost parallel gullies in the walls, and subsidiary valleys on their floors which are slightly converging down slope. This figure shows canyons at a more advanced stage of development than those illustrated in Fig. 95. View as obliquely as possible from the bottom of the page for the most realistic effect.

Continental slope off Algeria, Mediterranean.   36°17′N 00°28′E
Minimum depth on the continental shelf 115 m.   W.E. × 4
Maximum depth in canyons about 900 m.

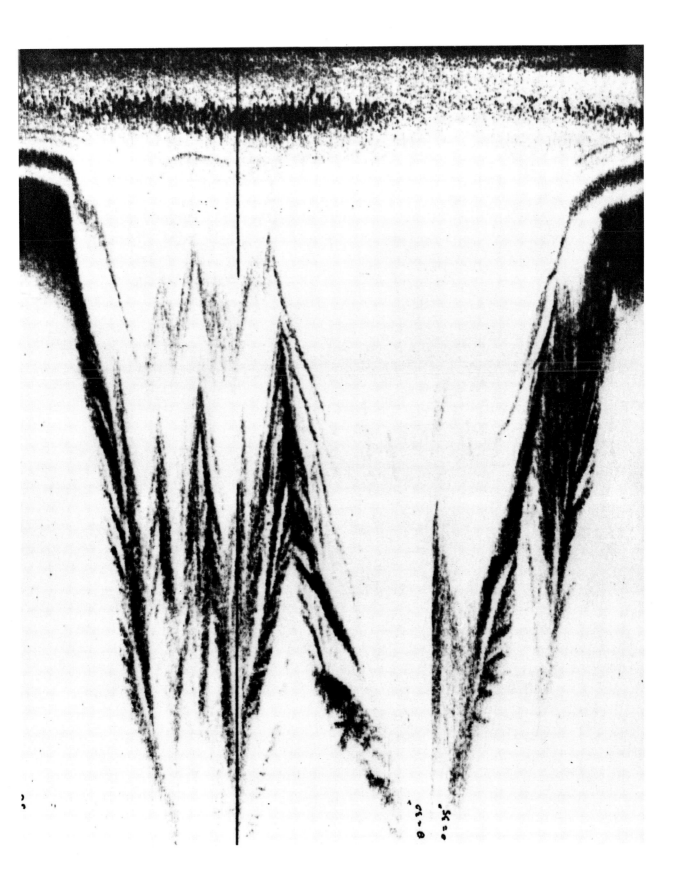

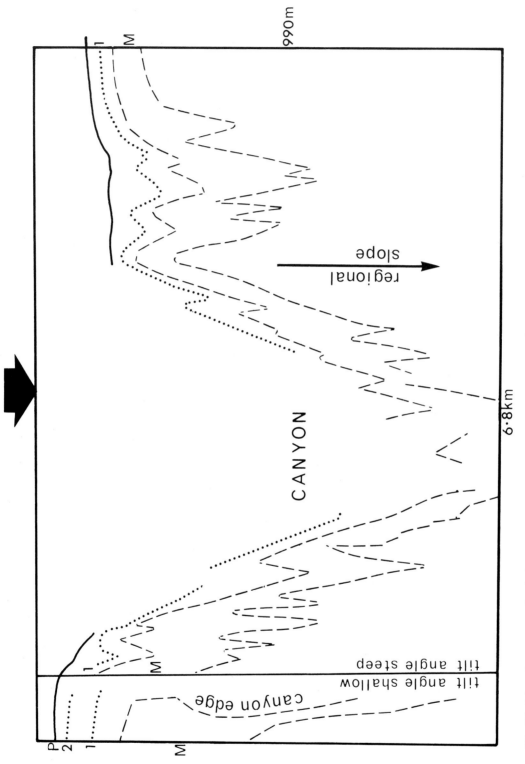

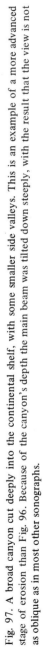

Fig. 97. A broad canyon cut deeply into the continental shelf, with some smaller side valleys. This is an example of a more advanced stage of erosion than Fig. 96. Because of the canyon's depth the main beam was tilted down steeply, with the result that the view is not as oblique as in most other sonographs.

Continental slope off Algeria, Mediterranean.
Minimum depth on continental shelf 75 m.
Maximum depth in canyon about 950 m.

36°30'N   00°59'E
W.E. × 4

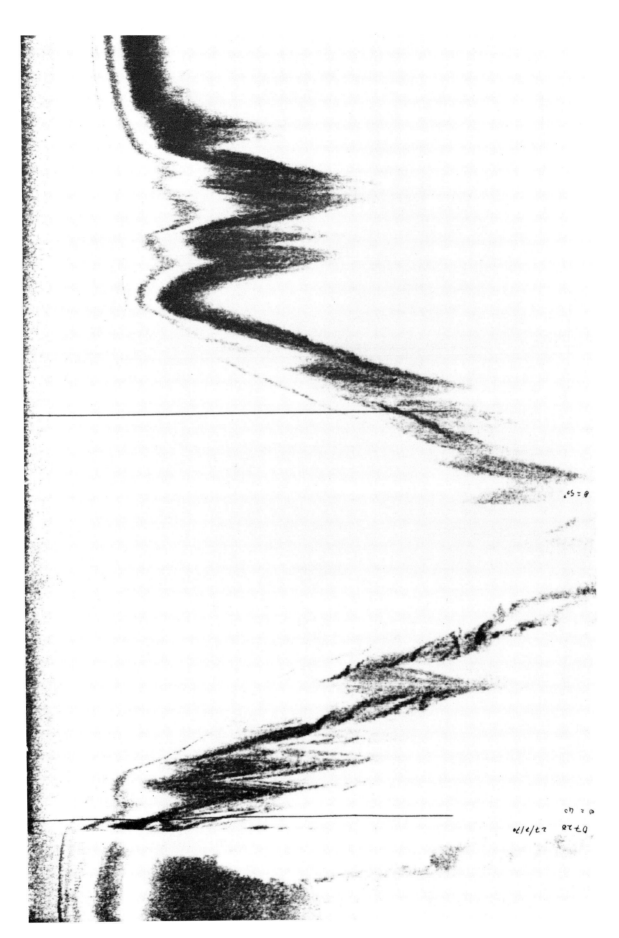

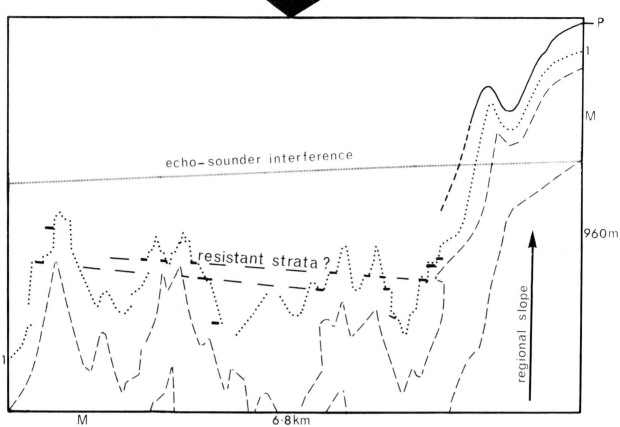

Fig. 98. A view looking up a canyoned slope, one main canyon wall of which is shown. The slope is remarkable for the pagoda-like cross-section of the ridges between the three main valleys. The ledges on these ridges are probably due to the outcrop of two or perhaps three relatively resistant layers of rock, whose continuity is most evident if the figure is viewed obliquely from the right hand side. In the deeper water the profile P is missing, but echoes from side lobe 1 serve as a useful substitute.

Continental slope off Algeria, Mediterranean.   36°18′N 00°30′E
Depth on continental shelf about 230 m.   W.E. ✕ 4
Greatest depth in canyon about 950 m.

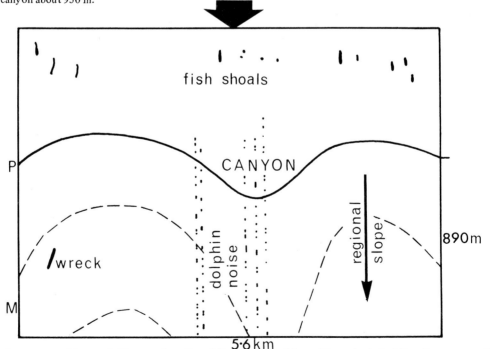

Fig. 99. A submarine canyon with remarkably smooth walls and floor, devoid of the gullies so typical of the other canyons illustrated in this book. Its smoothness is thought to result from deposition. The wreck (black blob) casts a shadow (white) beyond it.

Continental slope, west of Portugal.   39°34′N 09°37′W
Minimum water depth 360 m.   W.E. ✕ 2.5

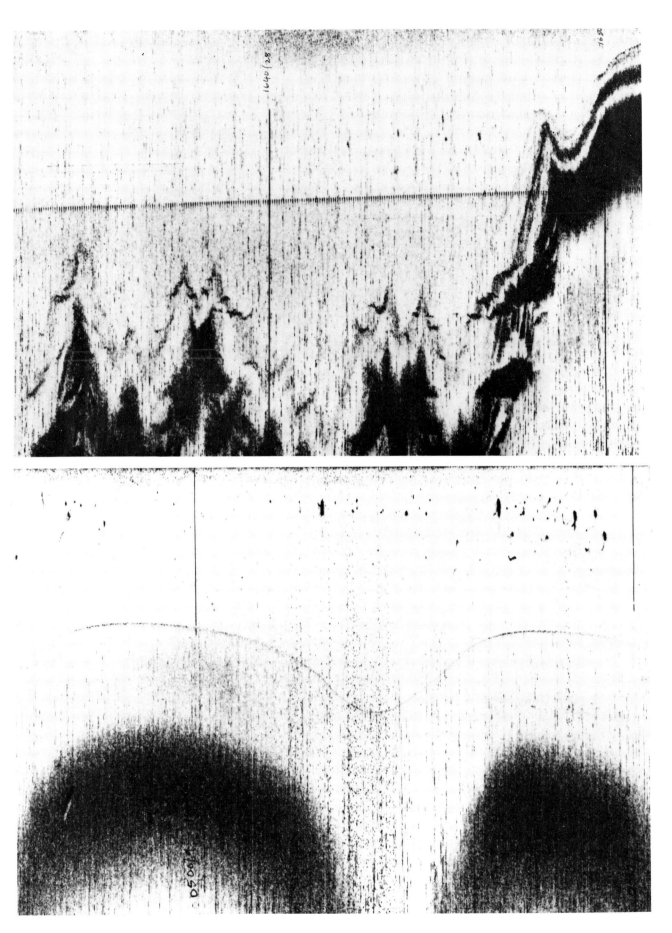

Cliffs

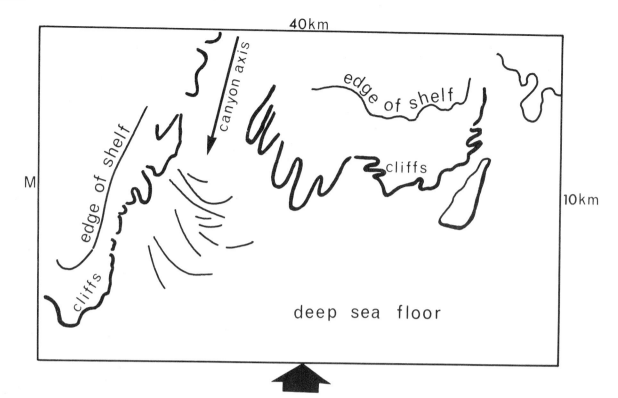

Fig. 100. Smooth deep-lying floor which gives way abruptly to stepped cliffs (near the isle of Capri) with an average slope of 15° and a height of about 800 m. The cliffs are interrupted by a submarine canyon, issuing from the Bay of Naples, on whose sediment-covered floor there are arcuate features interpreted as low ridges. These may result from sediment moving down canyon or possibly down the gullies in the cliff on the left hand side of the sonograph. The same information is presented on both sonographs, but at different width scales.

Continental slope, Tyrrhenian Sea, Mediterranean.
Water depth of deep floor about 1,000 m.
GLORIA

40°30′N 14°05′E
W.E. top × 2.5
W.E. bottom × 1

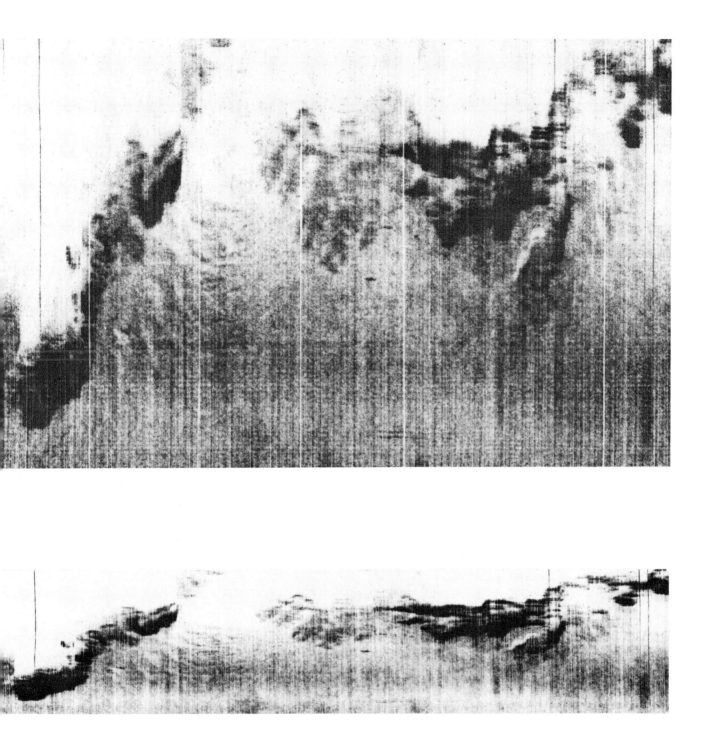

Ridges

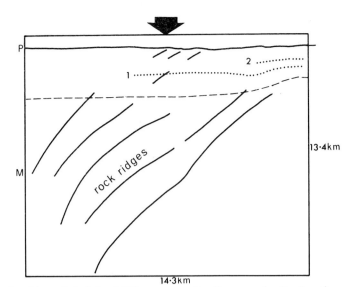

Fig. 101. Parallel ridges and troughs with a relief of about 150 m, representing the seaward extension of a young fold belt. The somewhat stepped outline of the ridges is due to a mechanical fault in the equipment.

Continental slope, Gulf of Cadiz.   36°15′N 07°25′W
Water depth in troughs about 1,000 m.   W.E. X 1
GLORIA

Submarine Canyons

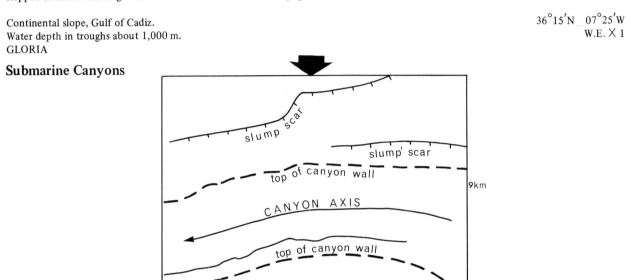

Fig. 102. Part of a submarine canyon, which is up to 4.5 km wide and 650 m deep near the top of the continental slope, but only 3.5 km wide lower down. It is probable that this narrowing results partly from the collapse of one wall into the canyon, as large slumps are indicated by a sub-bottom profile close to the edge of this canyon. Much of the canyon is in shadow (white).

Continental slope, off eastern Spain.   41°15′N 02°35′E
Water depth of continental slope about 400 to 1,100 m.   W.E. X 1
GLORIA

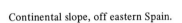

Fig. 103. A view up the continental slope showing narrow, sharply defined bands of relatively rough floor (dark tone) ½ to 1 km wide, some of which converge with one another towards the bottom of the continental slope. Echo-sounding traverses over the ground show the presence of submarine cayons up to 150 m deep, while Fig. 92 shows more detail near the top of this slope. The overall gradient is about 9°

Continental slope off Algiers, Mediterranean.   37°00 N′ 03°40′E
Water depth at foot of continental slope about 2,250 m.   W.E. X 2.5
GLORIA

Submarine Canyons

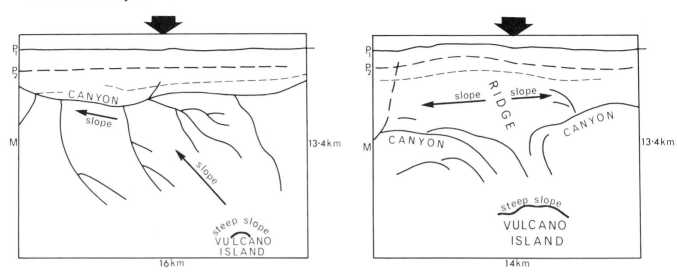

Fig. 104. A group of tributary canyons on the submarine slopes of Vulcano Island, seen in two overlapping sonographs oriented about 90° to each other, which are presented with expanded scale at the top (W.E. × 2.5) and almost true plan view at the bottom of the page. These small canyons are up to 50 m deep, while the larger one (Stromboli Canyon), into which some of them lead, is about 250 m deep. For each channel the side facing the illumination direction gives a strong reflection and so appears as a black line while the other side is in shadow and so appears as a white line. Soon after an earthquake in 1908 a telephone cable which crosses the main canyon was broken, presumably by a turbidity current. P_2 is a multiple reflection of bottom profile P_1.

Aeolian Islands, Tyrrhenian Sea, Mediterranean. 38°25'N 15°00'E
Maximum water depth about 1,170 m. W.E. top × 2.5
GLORIA W.E. bottom × 1

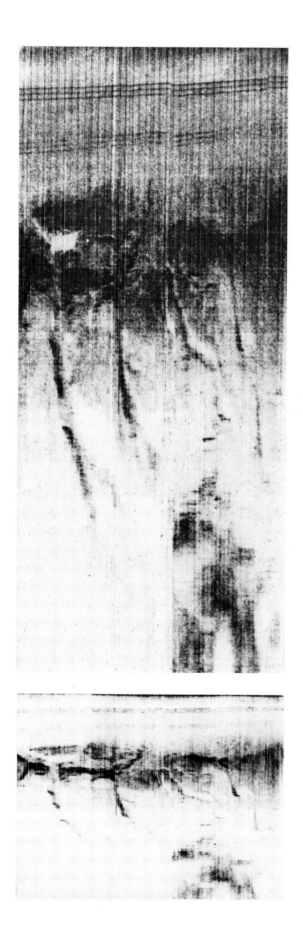
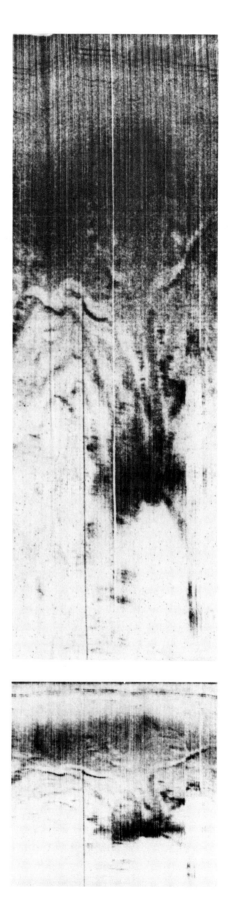

Seamount

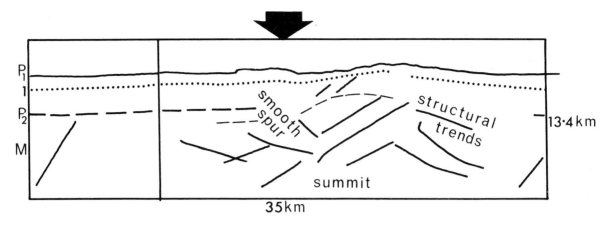

Fig. 105. One half of a large magnetic seamount (Marsili Seamount), which rises about 1000 m in profile P and is known to be about 700 m higher towards the bottom of the figure. In profile the seamount is asymmetrical and broken into a number of steps, while in plan view the relief follows two linear trends, at about 70° to one another. These trends are interpreted as indicating the position of two sets of faults. Alongside there is a relatively smooth, but steep-edged spur with a height of about 250 m. Surfaces facing the illumination direction appear black with white shadows behind them. The same information is presented on both sonographs, but at different width scales. P_2 is a multiple reflection of bottom profile P_1.

39°19′N 14°23′E
W.E. top × 2
W.E. bottom × 1

Tyrrhenian Sea, Mediterranean.
Maximum water depth about 3,330 m.
GLORIA

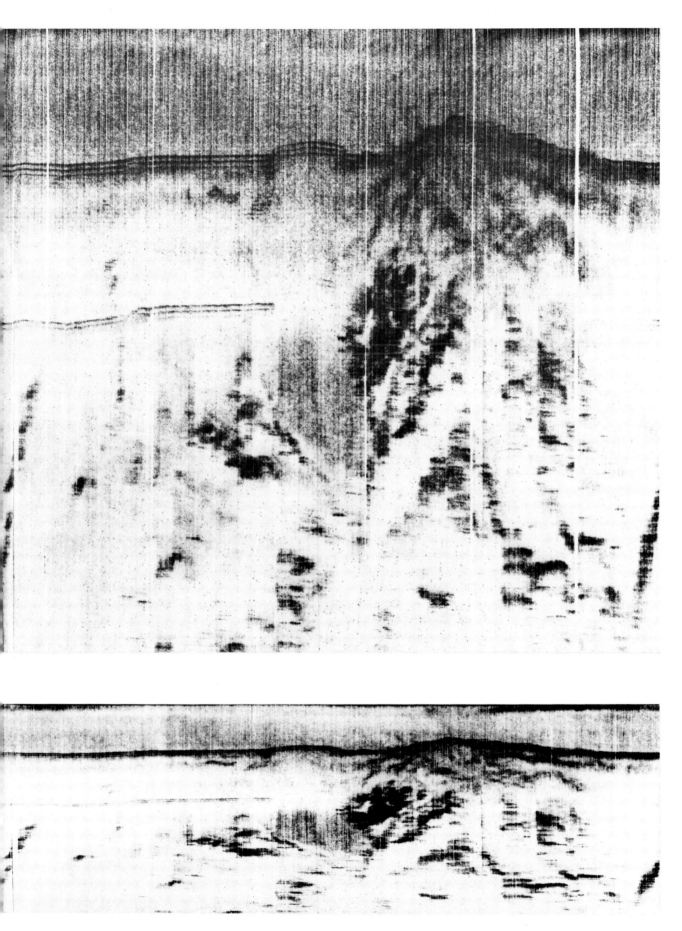

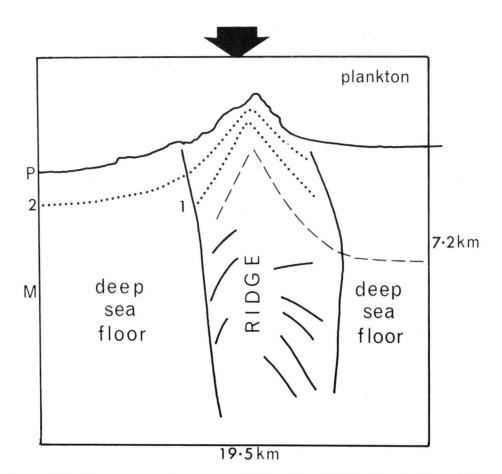

Fig. 106. The rough crestal ridge of a non-magnetic, linear seamount, 11 km wide and 2,000 m high, stands in marked contrast to the smooth depositional floors, on one flank at 2,000 m and the other at 2,900 m. Surfaces facing the ship appear black, while shadows are white.

Tyrrhenian Sea, Mediterranean.   40°48'N   11°03'E
Minimum water depth 980 m.   W.E. × 2
GLORIA

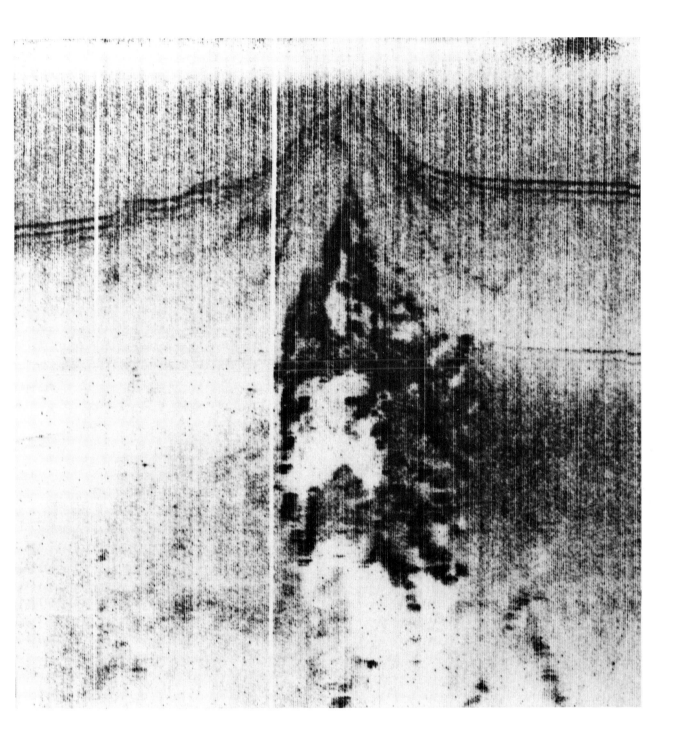

Deep-sea Channel

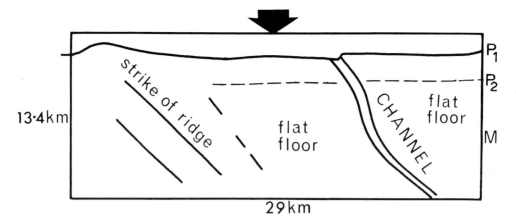

Fig. 107. A 9-km length of a deep-sea channel, about 1 km wide, which reaches a depth of more than 130 m below the flat depositional floors around it. The channel, which is heading from the foot of a canyon east of Sardinia, may follow the line of a fault, whose presence is suggested by the 100 m difference in depth on either side of the channel. Surfaces facing the ship appear black, while shadows are white. P_2 is a multiple reflection of bottom profile P_1.

Tyrrhenian Sea, Mediterranean.   40°15'N 10°20'E
Water depth of flat floor about 2,100 m.   Sonograph W.E. × 2
GLORIA   Diagram W.E × 1

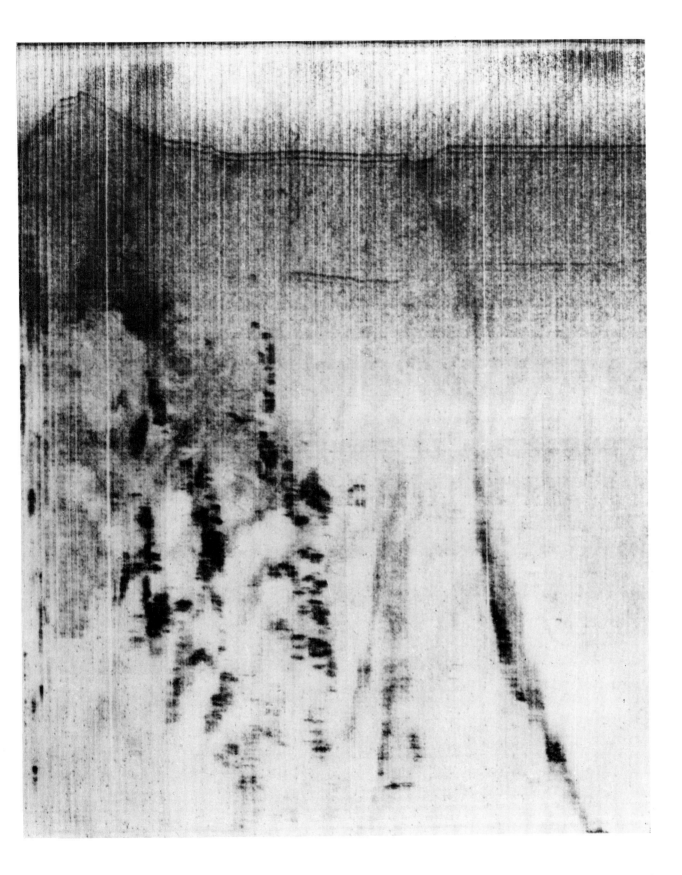

125

Volcanoes

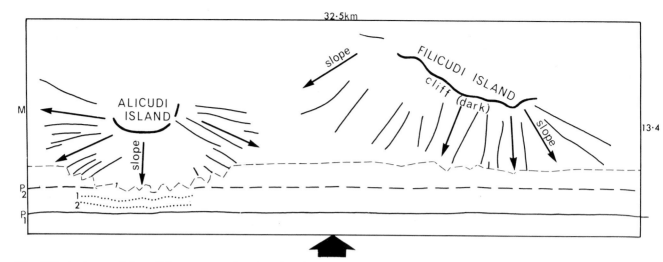

Fig. 108. A characteristic radial pattern on the submarine slopes of recently active volcanic islands. The pattern is interpreted as indicating lava flows, ash slides or else gullies resulting from erosion. There are relatively steep slopes (dark bands) around these volcanoes at a depth of between 200 and 300 m, which may indicate a submerged line of cliffs. Ground above sea level and beyond the islands appears white. The same information is presented on both sonographs, but at different width scales. P_2 is a multiple reflection of bottom profile P_1.

Aeolian Islands, Tyrrhenian Sea, Mediterranean.
Water depth about 1,500 m.
GLORIA

38°30'N 14°30'E
W.E. top × 2.5
W.E. bottom × 1

Tectonic Relief

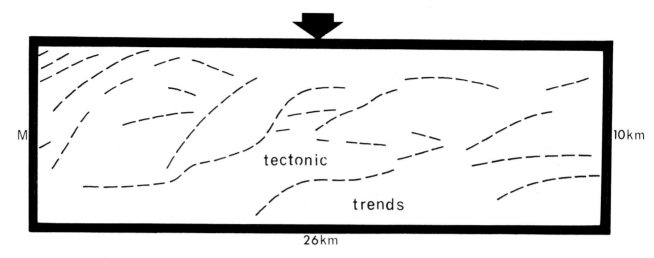

Fig. 109. Young tectonic relief, up to about 70 m high, on the Mediterranean Ridge, formed in soft sediments by squashing between Europe and Africa. View the sonographs obliquely from the bottom at a distance of about 2 m. The relief then stands out, with shadows in black (negative print). The same information is presented on both sonographs, but at different width scales.

Eastern Mediterranean.   33°43′N   23°05′E
Water depth about 800 m.   W.E. top × 2.5
GLORIA   W.E. bottom × 1

II. OTHER SONOGRAPHS

Fish Shoals

Echoes from fish shoals can be recognised with certainty when they are wholly or partly recorded above the profile *P* or between the side lobes. This helps to distinguish between fish shoals and sea floor features where these are recorded together.

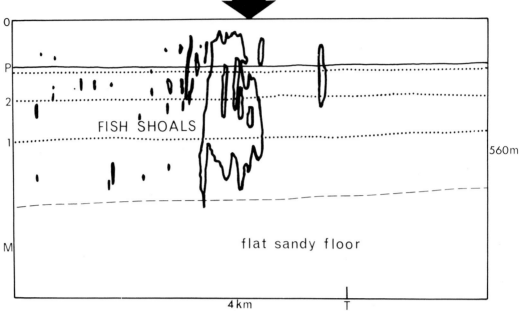

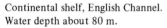

Fig. 110. Close spaced fish shoals extend across the side lobe echoes.

Continental shelf, English Channel.
Water depth about 80 m.

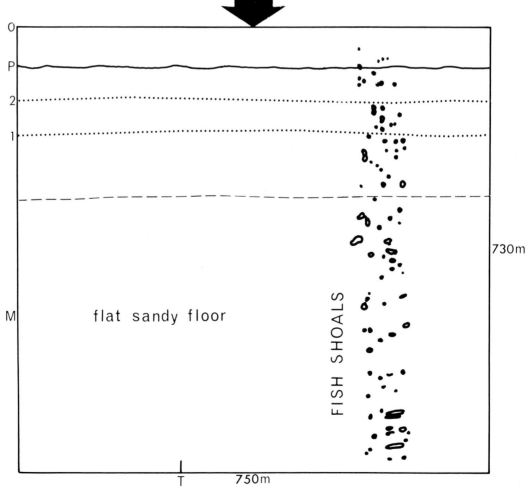

Fig. 111. A large number of small fish targets are distributed throughout the sonograph.

Continental shelf, English Channel.
Water depth about 68 m.

132

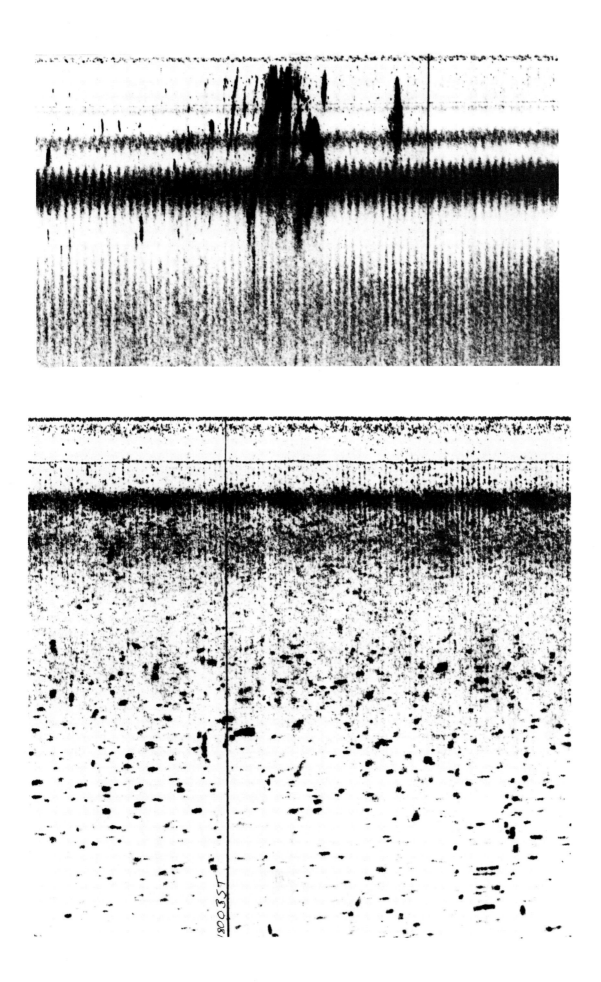

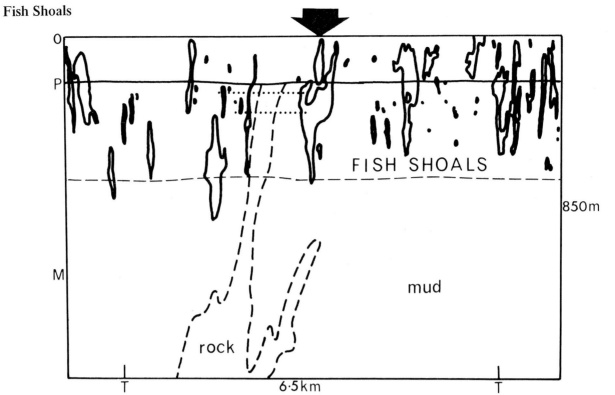

Fig. 112. Most of these fish shoals are easily recognised, but confusion arises when their echoes are recorded together with sea floor patterns.

Continental shelf, west of Portugal.   38°55′N 09°40′W
Water depth about 120 m.   W.E. × 5

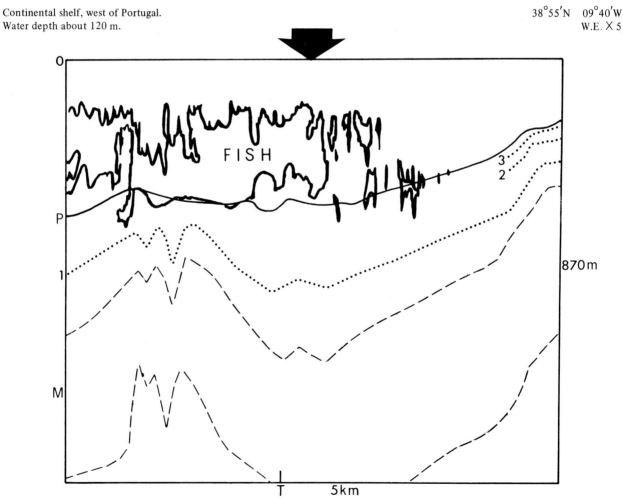

Fig. 113. An extensive layer of fish shoals over deep water. The side lobes can be used to follow the topography out sideways from beneath the ship.

Continental shelf, Tyrrhenian Sea, Mediterranean.   38°22′N 15°34′E
Maximum water depth beneath ship about 350 m.   W.E. × 5

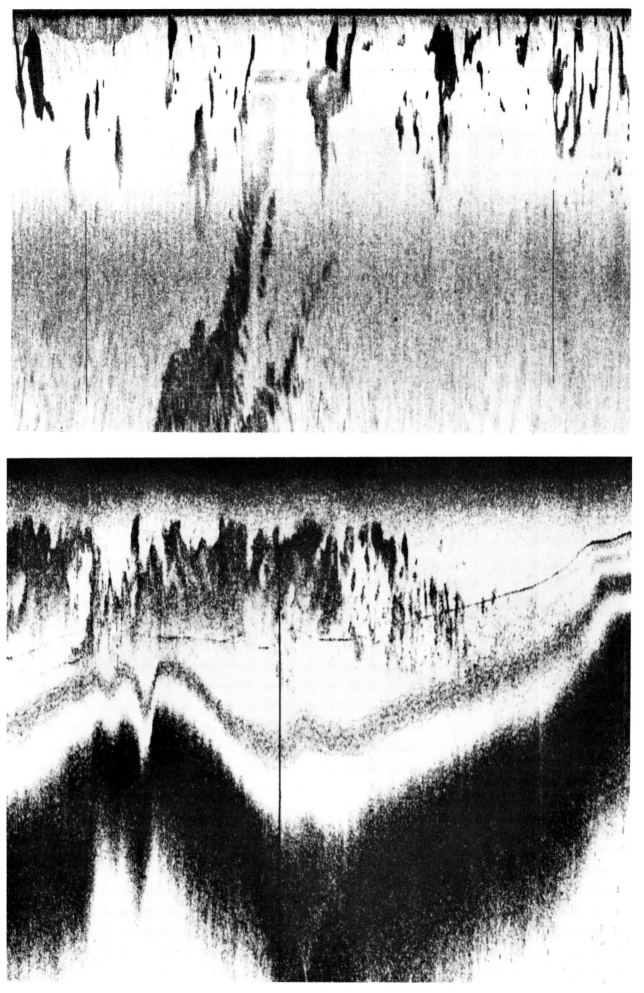

Fish Shadows

As sound passes through a fish shoal some energy is absorbed and some is lost by reflection. When the fish density is large enough a recognisable shadow will be cast on to the sea floor.

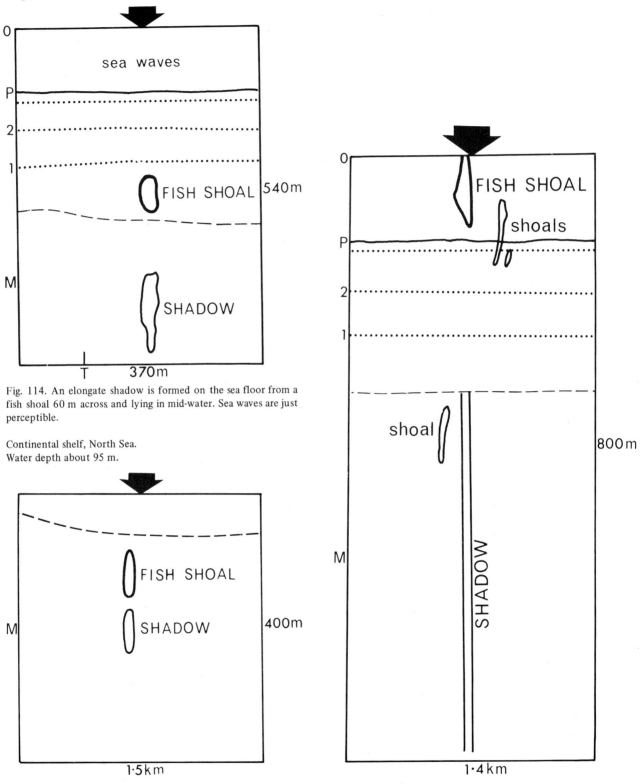

Fig. 114. An elongate shadow is formed on the sea floor from a fish shoal 60 m across and lying in mid-water. Sea waves are just perceptible.

Continental shelf, North Sea.
Water depth about 95 m.

Fig. 115. A shadow is produced by a fish shoal about 60 m across and lying 30 m above the floor in deep water. Diffuse sea bed features are also present.

Continental shelf, Mediterranean.
Water depth about 340 m.

Fig. 116. A fish shoal about 80 m across and very close to the transducer reduces the strength of the sea bed echoes across the full width of the sonograph.

Continental shelf, North Sea.
Water depth about 130 m.

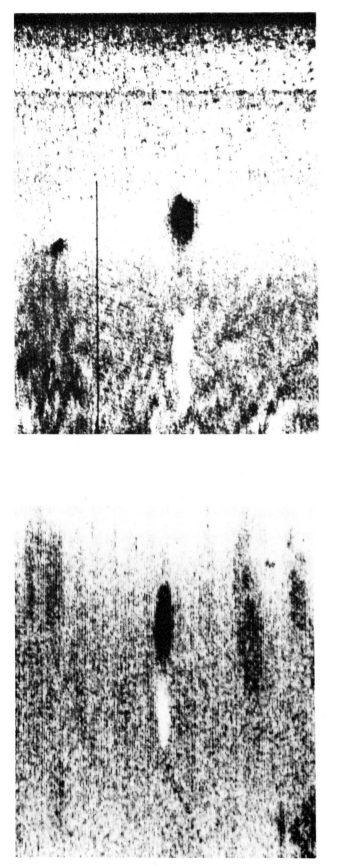

shadow from fish school

Scattering Layers

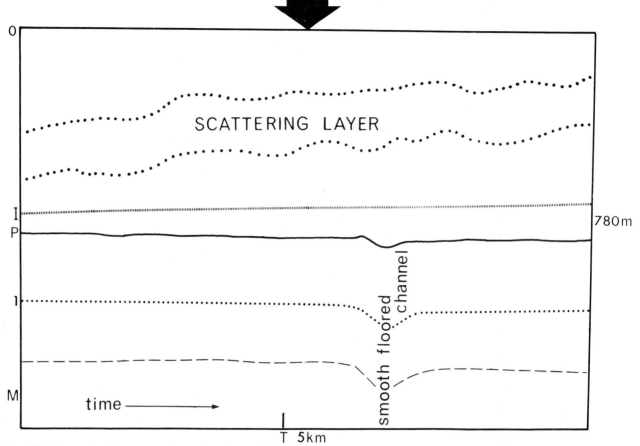

Fig. 117. A layer of midwater organisms undergoing vertical migration at sunset. Other evidence is required to determine the proportions of plankton to fish. *I* indicates echo-sounder interference.

Continental shelf, Gulf of Cadiz.
Water depth about 400 m.

36°22'N   06°48'W

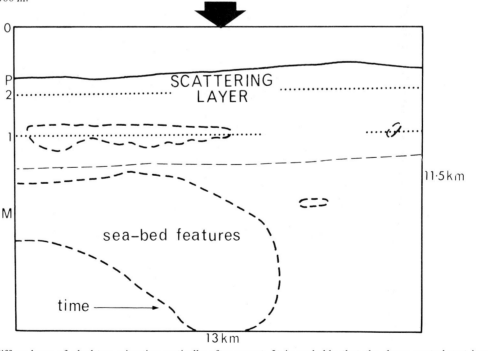

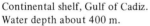

Fig. 118. A large diffuse layer of plankton migrating vertically after sunset. It is probable that the dense central part is caused by resonant effects of the plankton gas bubbles.

Tyrrhenian Sea, Mediterranean.
Maximum water depth about 2,200 m.
GLORIA

39°25'N   13°25'E
W.E. × 1

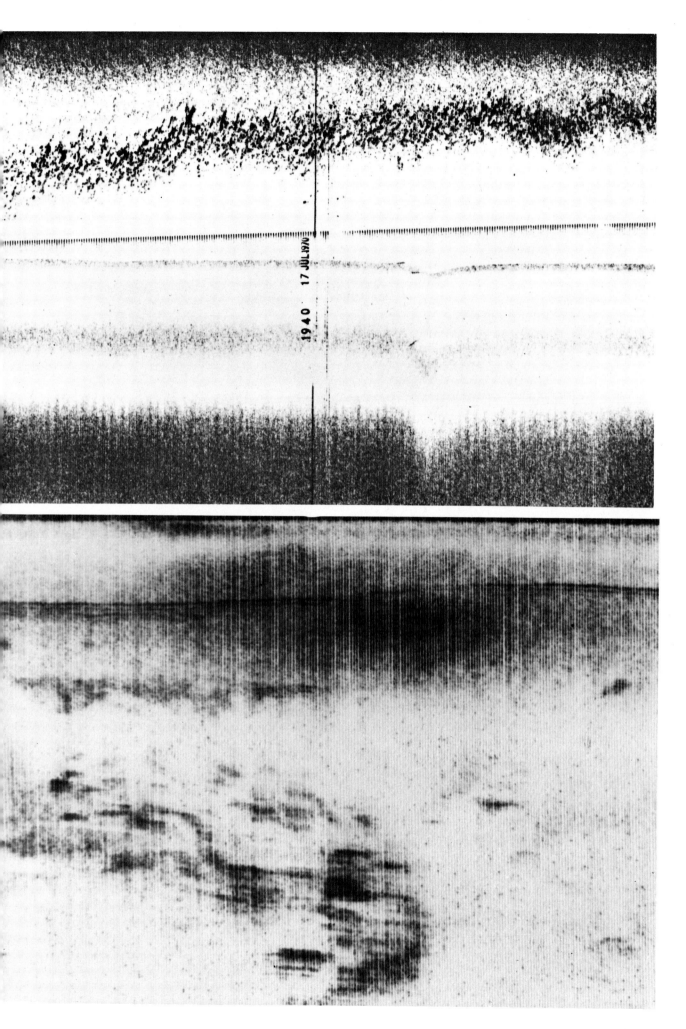

17 JUL 1970

1940

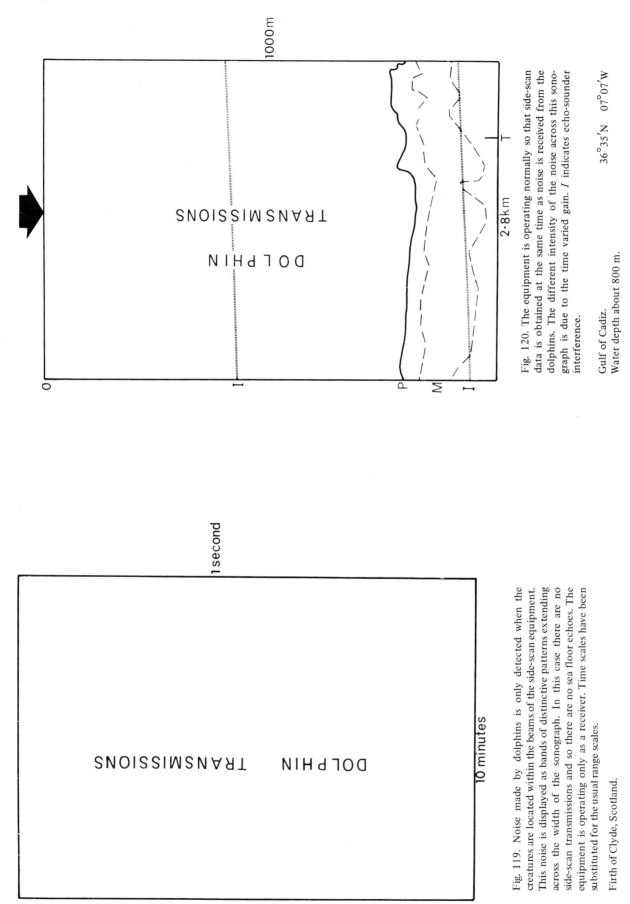

Fig. 120. The equipment is operating normally so that side-scan data is obtained at the same time as noise is received from the dolphins. The different intensity of the noise across this sonograph is due to the time varied gain. *I* indicates echo-sounder interference.

Gulf of Cadiz.   36°35′N   07°07′W
Water depth about 800 m.

Fig. 119. Noise made by dolphins is only detected when the creatures are located within the beams of the side-scan equipment. This noise is displayed as bands of distinctive patterns extending across the width of the sonograph. In this case there are no side-scan transmissions and so there are no sea floor echoes. The equipment is operating only as a receiver. Time scales have been substituted for the usual range scales.

Firth of Clyde, Scotland.

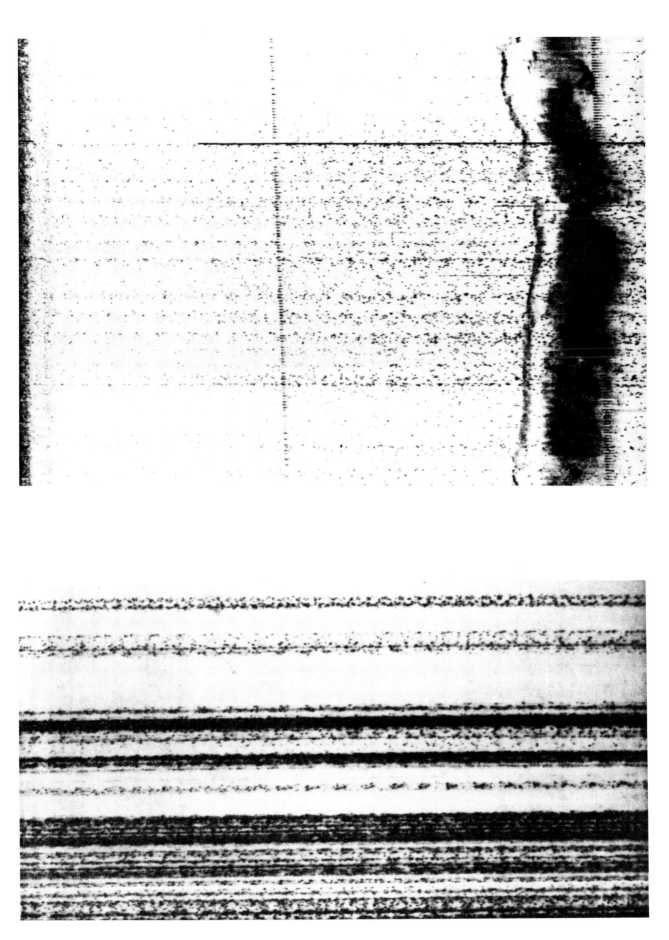

Sea Waves

Waves at the sea surface can give rise to characteristic patterns particularly when the main beam reaches the sea surface.

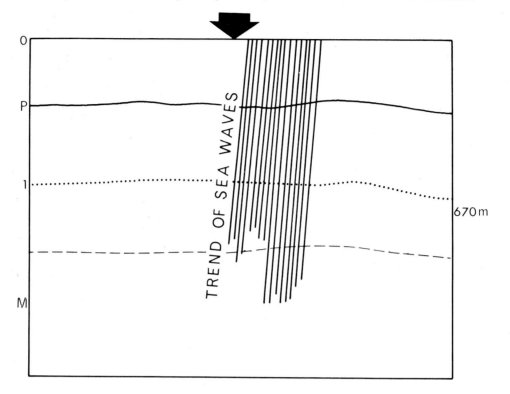

Fig. 121. A well-developed wave pattern partially obscures echoes from the sea floor.

Continental shelf, North Sea.
Water depth about 130 m.

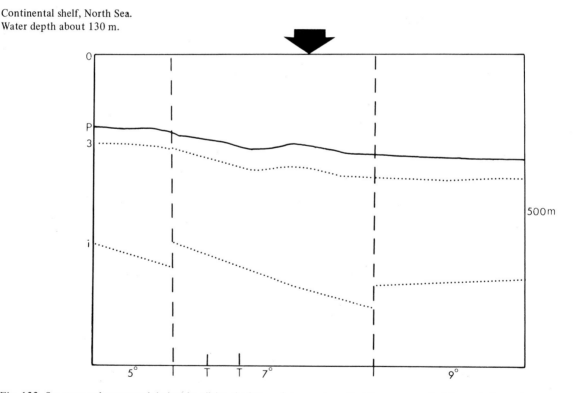

Fig. 122. Sea wave echoes are minimised by tilting the beams downwards so that the upper part of the main beam does not reach the sea surface. Note the steps in the side lobe echoes as the angle is changed.

Continental shelf, English Channel.
Maximum water depth about 170 m.

142

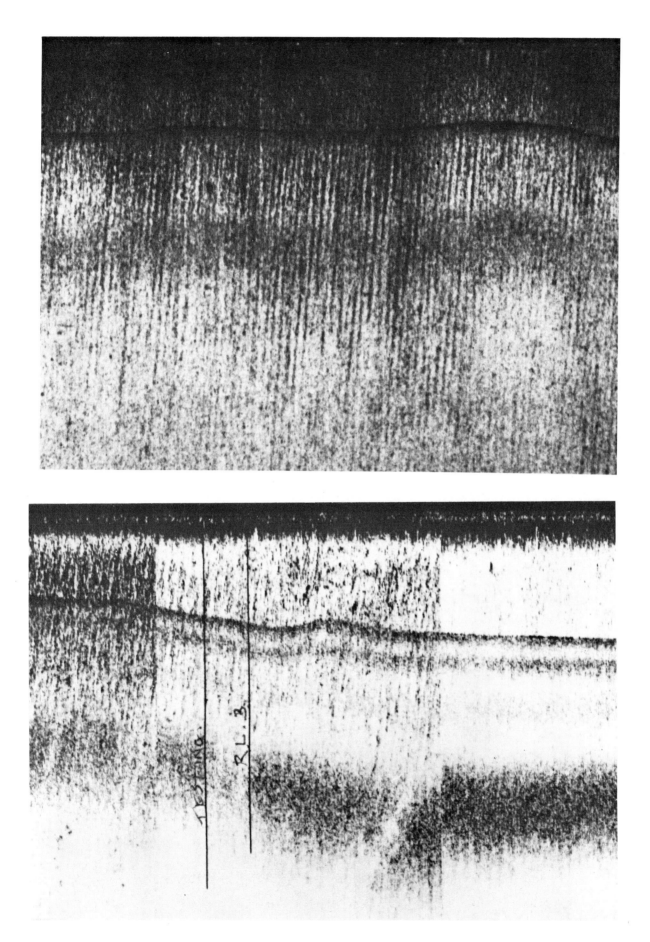

Turbulence

Strong turbulence at the sea surface can cause air bubbles to be drawn into the water, giving rise to two unwanted effects on sonographs. Echoes can be obtained from the bubbles and in addition the bubbles themselves may act as noise sources. Noise has also been observed emanating from isolated points on sand wave crests, where it probably results from sand movement.

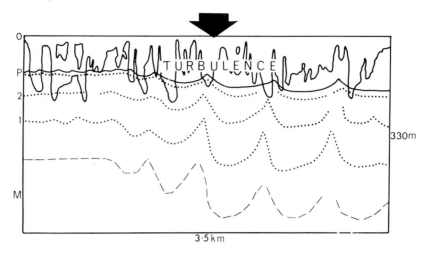

Fig. 123. Turbulence produced by strong tidal currents over a ridged sea floor is thought to be responsible for the echoes at the top of the sonograph. The sand or gravel ridges are up to 25 m high.

Continental shelf, Irish Sea.
Maximum water depth about 90 m.

52°48′N   05°32′W
W.E. × 5

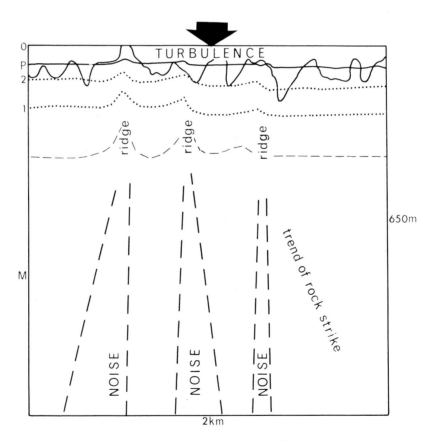

Fig. 124. Noise generated as water passes over three rock ridges is radiated in all directions but only received when the narrow sound beam of the side-scan sonar pointed towards a source. The bands appear triangular because the amplifier gain increases with range.

Continental shelf, Bristol Channel.
Maximum water depth about 33 m.

51°13′N   04°13′W

144

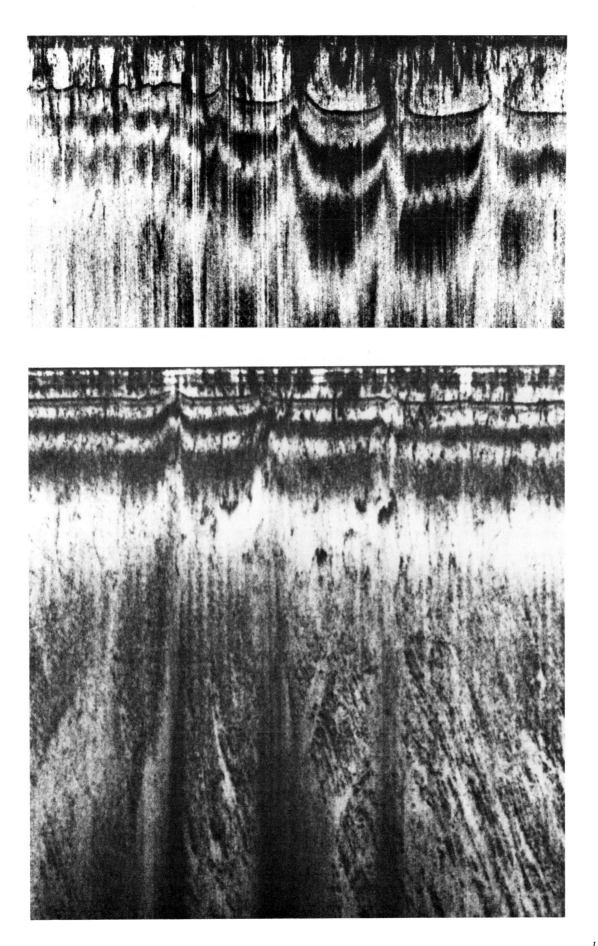

Quenching

When a ship is pitching or rolling, or waves are breaking, air bubbles trapped in the water can form a temporary screen across the front of a transducer so that no useful information can be obtained with the equipment. Such periods are represented by white bands extending across the sonograph. Wakes of vessels can also serve as effective bubble screens.

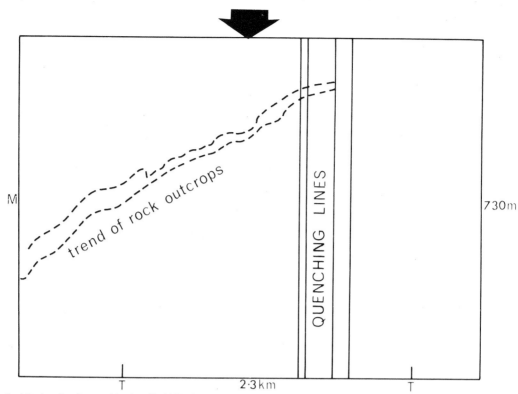

Fig. 125. Broad white bands of quenching locally hide the pattern of rock outcrops, although the general trend is still evident.

Continental shelf, English Channel.
Water depth about 60 m.

50°16′N _ 01°30′W
W̄.E. × 2.5

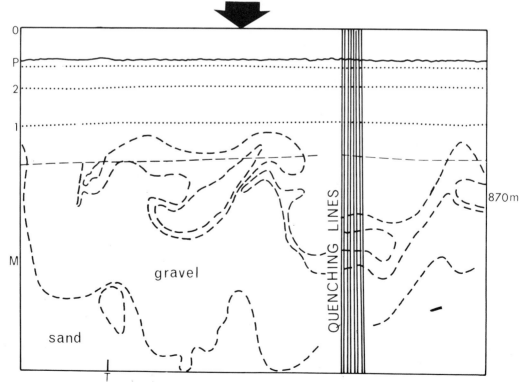

Fig. 126. Trends of sea floor features are more obscured when parallel to the bands of quenching than when at right angles to them.

Continental shelf, English Channel.
Water depth about 90 m.

146

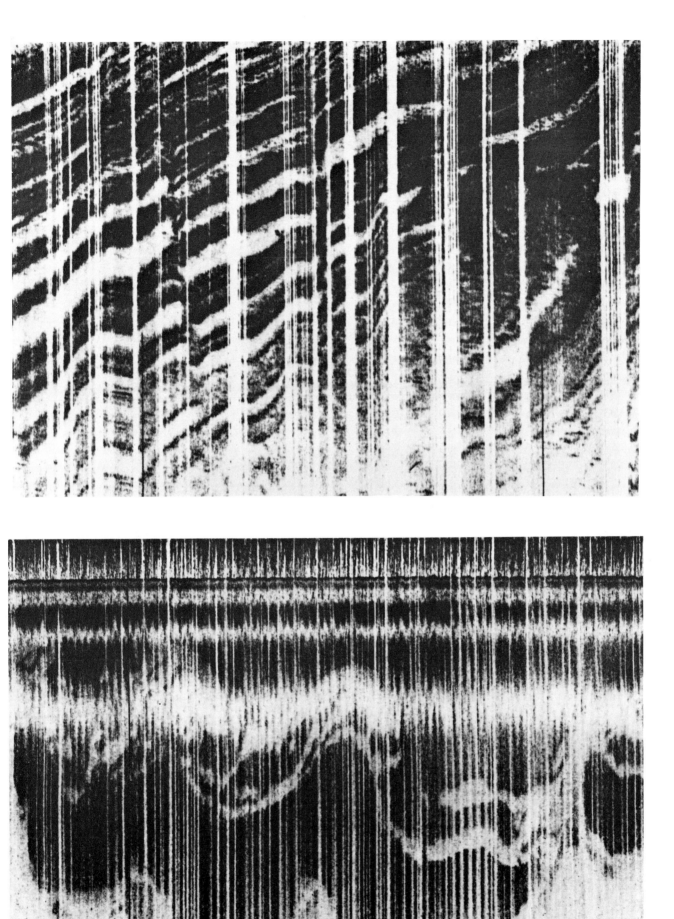

Lloyd Mirror

Lloyd Mirror bands usually extend along the length of a sonograph. Although sea floor details tend to be obscured, the bands can be of value for emphasising the overall relief.

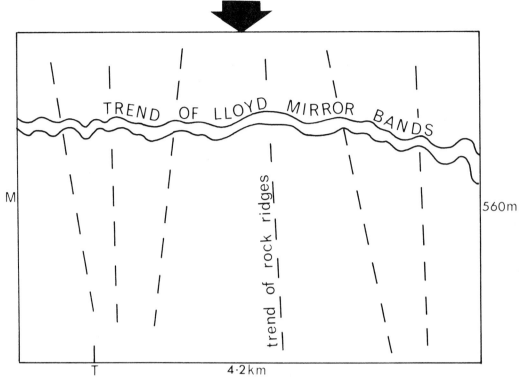

Fig. 127. Undulations in the Lloyd Mirror bands show that the outcrops of hard layers of rock form broad, low ridges. The pattern of dots on the lower half of the sonograph is due to echo-sounder interference.

Continental shelf, Bristol Channel.
Water depth about 15 m.

51°24′N 03°37′W
W.E. X 5

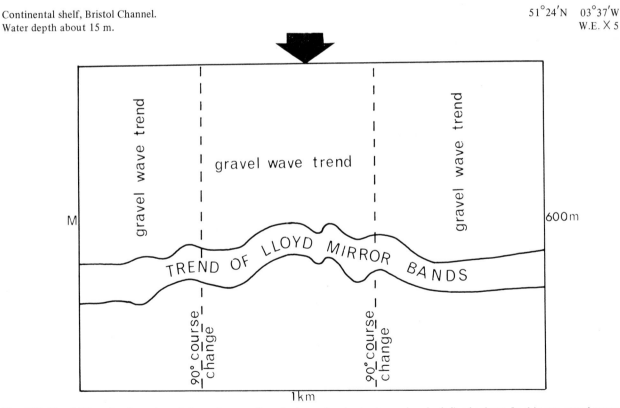

Fig. 128. Lloyd Mirror bands are less obvious where sea floor features give rise to strong sharply defined echoes. In this case gravel waves are clearly visible when parallel to the ship's course, whereas the Lloyd Mirror bands predominate when the gravel wave crests and the ship's course are at right angles.

Continental shelf, Bristol Channel.
Water depth about 10 m.

51°15′N 03°41′W
W.E. X 1

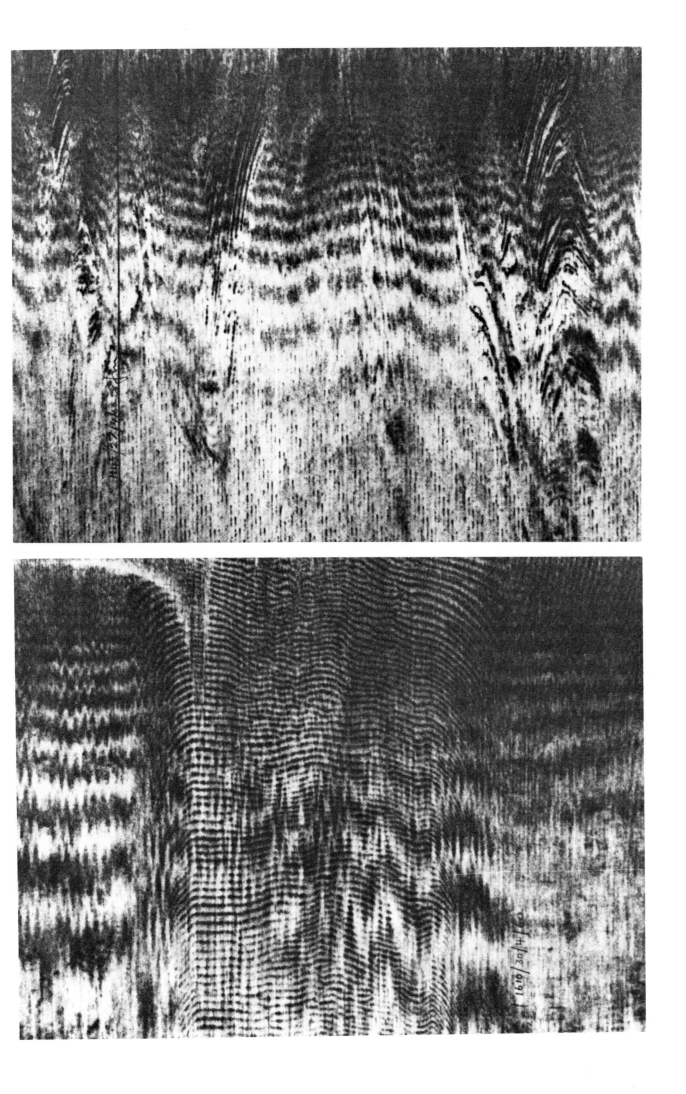

Lloyd Mirror

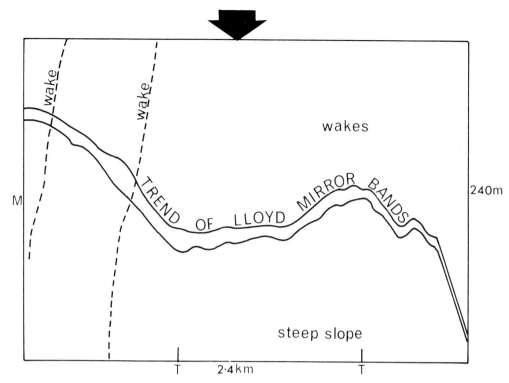

Fig. 129. For smooth sea floors Lloyd Mirror bands can be well defined. Here, undulations of the sea floor are clearly indicated by the bands in the vicinity of an island.

Continental shelf, Hong Kong.　　　　　　　　　　　　　　　　　　　　22°27′N　114°15′E
Maximum water depth 35 m.　　　　　　　　　　　　　　　　　　　　　　W.E. × 6
Towed Surveying Asdic. By courtesy of the Geophysics Group, Bath University.

Fig. 130. The relief on the edge of a shipping channel is emphasised by the Lloyd Mirror bands.

Continental shelf, Southampton Water, England.　　　　　　　　　　　50°43′N　01°29′W
Maximum water depth 50 m.　　　　　　　　　　　　　　　　　　　　　W.E. × 4.5

150

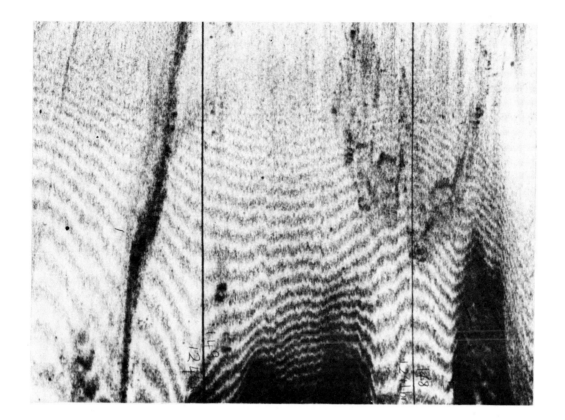

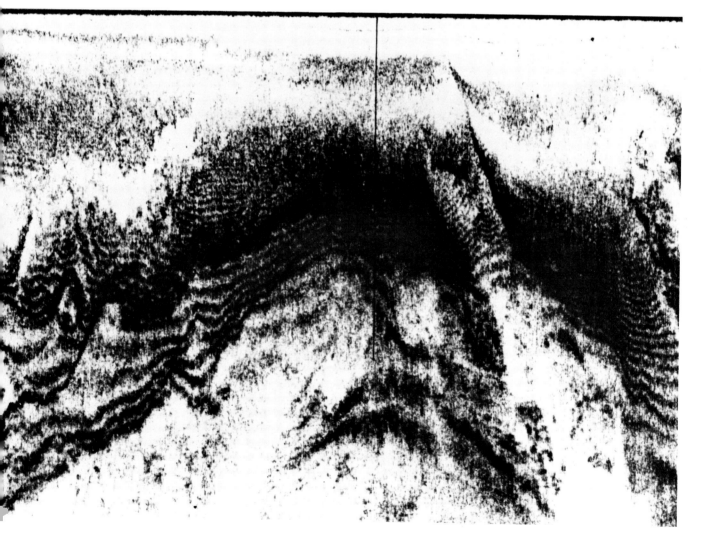

151

Sound Refraction

Density stratification, commonly due to temperature differences in the water, causes sound refraction which tends to limit the useful range. Complex refraction takes place when internal waves are also present and characteristic patterns are produced which can be confused with sea floor patterns. The effect varies with the incidence of storms, the locality and the time of year (referred to by month).

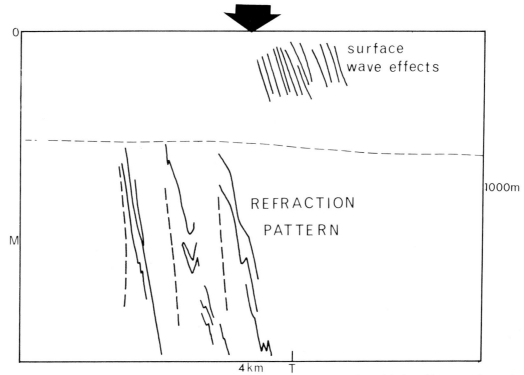

Fig. 131. Large scale temperature variations produce a well-developed refraction pattern that might be misinterpreted as rock outcrops.

Continental shelf, Gulf of Cadiz.
Water depth about 100 m.

36°34′N 06°42′W
October

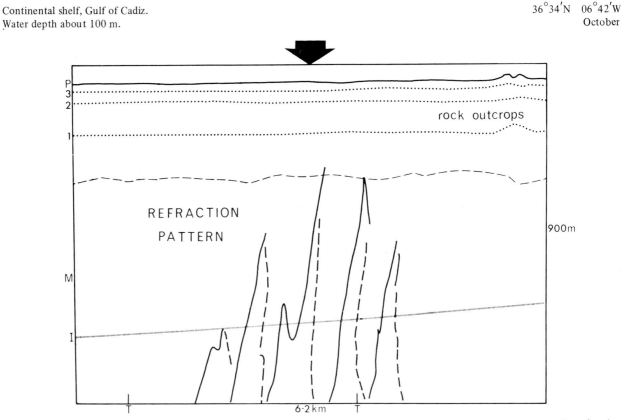

Fig. 132. A similar refraction pattern to that in Fig. 131, but here the sea floor character may be distinguished from the refraction pattern as the latter is not present in the side lobe echoes. *I* indicates echo-sounder interference.

Continental shelf, Celtic Sea.
Water depth about 130 m.

47°57′N 05°42′W
August

152

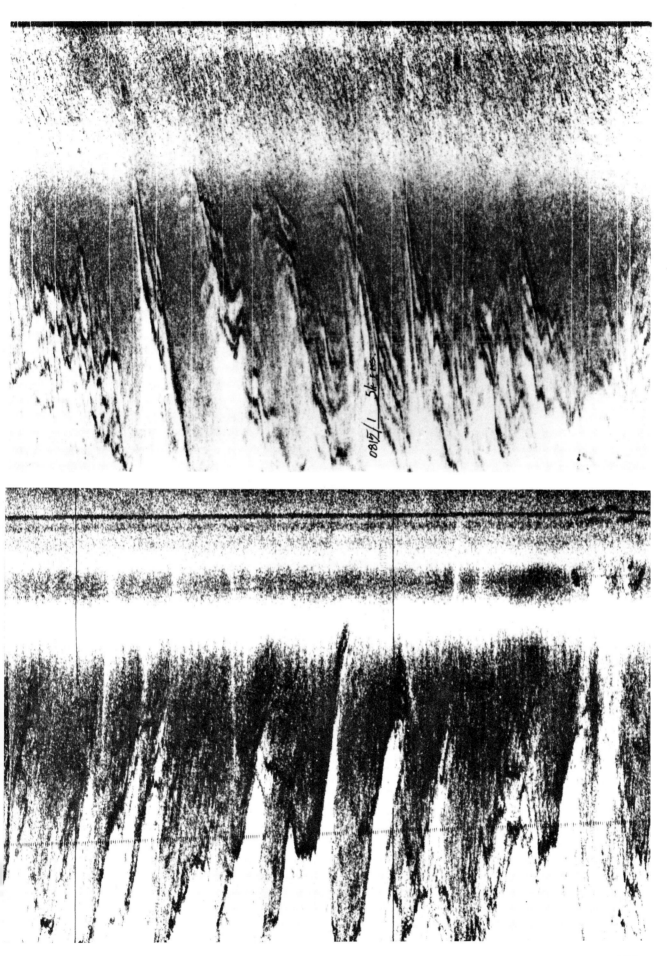

Sound Refraction

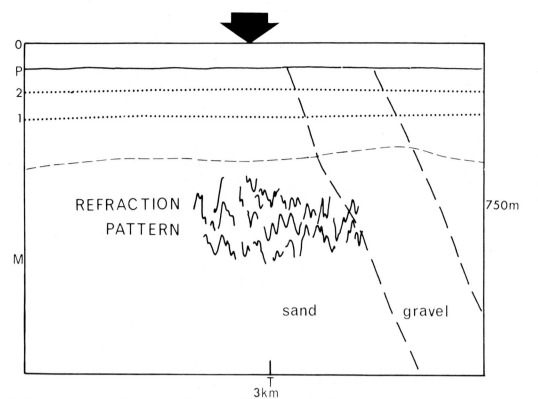

Fig. 133. Small-scale temperature variations produce the zig-zag refraction pattern which obscures sea floor detail. The gravel patch is still evident due to its higher reflectivity.

Continental shelf, Celtic Sea.   51°39'N 06°34'W
Water depth about 70 m.   May

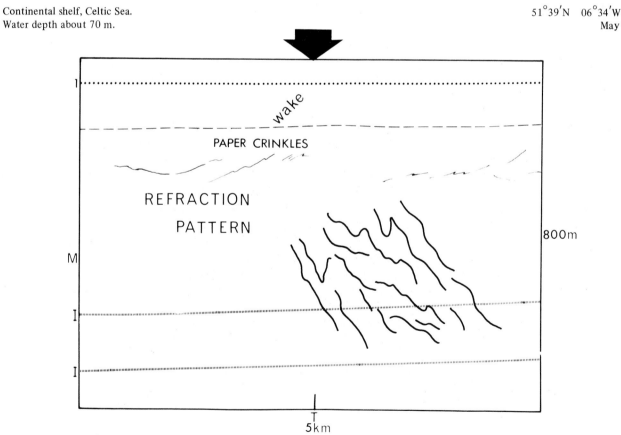

Fig. 134. A strongly-developed refraction pattern which completely obscures any features on the sea floor. *I* indicates echo-sounder interference.

Continental shelf, Mediterranean.   37°07'N 01°05'W
Water depth about 90 m.   July

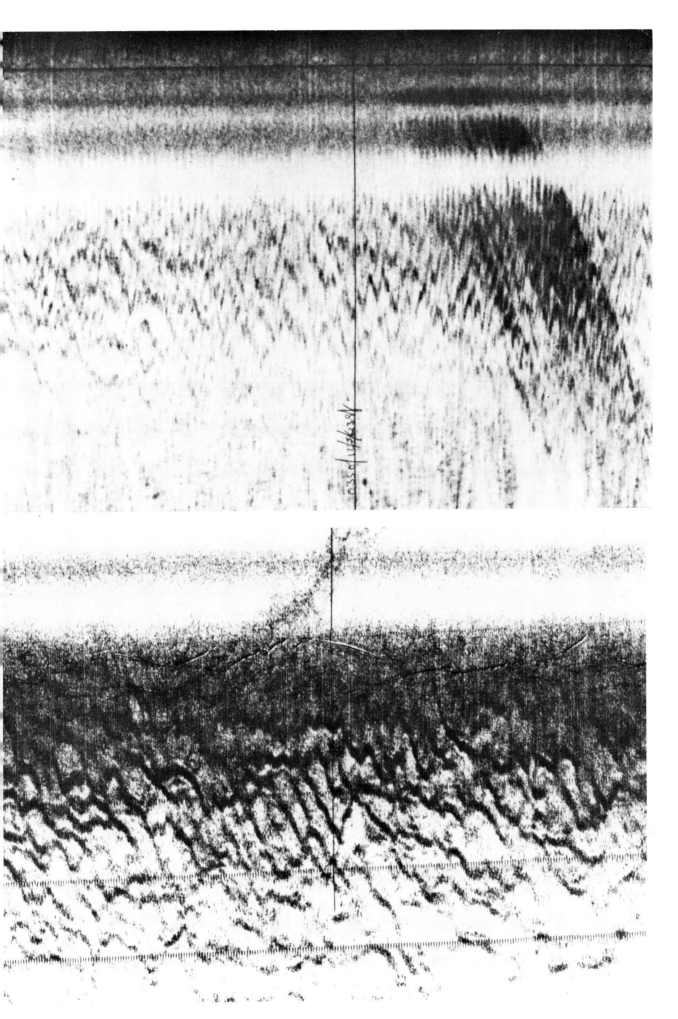

Sound Refraction

When there is a very sharp temperature gradient there may be sufficient refraction to produce surface or sea floor bouncing of the sound rays and then further refraction before sea floor echoes are returned.

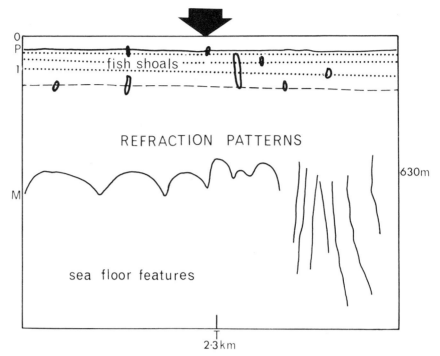

Fig. 135. A steep temperature gradient and an internal wave system have produced strong focussing. On the right of the sonograph the effect of the internal waves is more diffuse and so sea floor and refraction effects become indistinguishable.

Continental shelf, Gulf of Cadiz.   36°56′N   08°10′W
Water depth about 44 m.   September

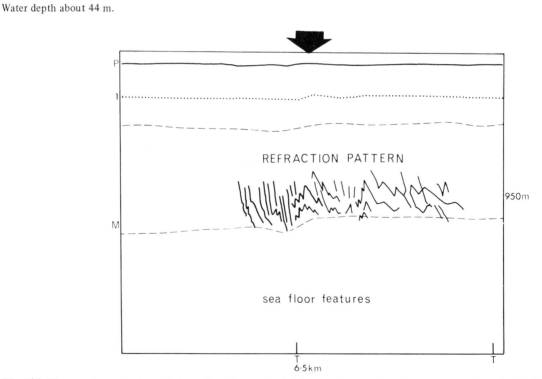

Fig. 136. The sound rays of the main beam have been refracted downward by a sharp temperature gradient thus limiting the maximum range. The sea floor echoes at the greater ranges are due either to bouncing of sound from the upper side lobe off the sea surface or of the main beam off the sea floor.

Continental shelf, Mediterranean.   41°16′N   09°39′E
Water depth about 90 m.   August

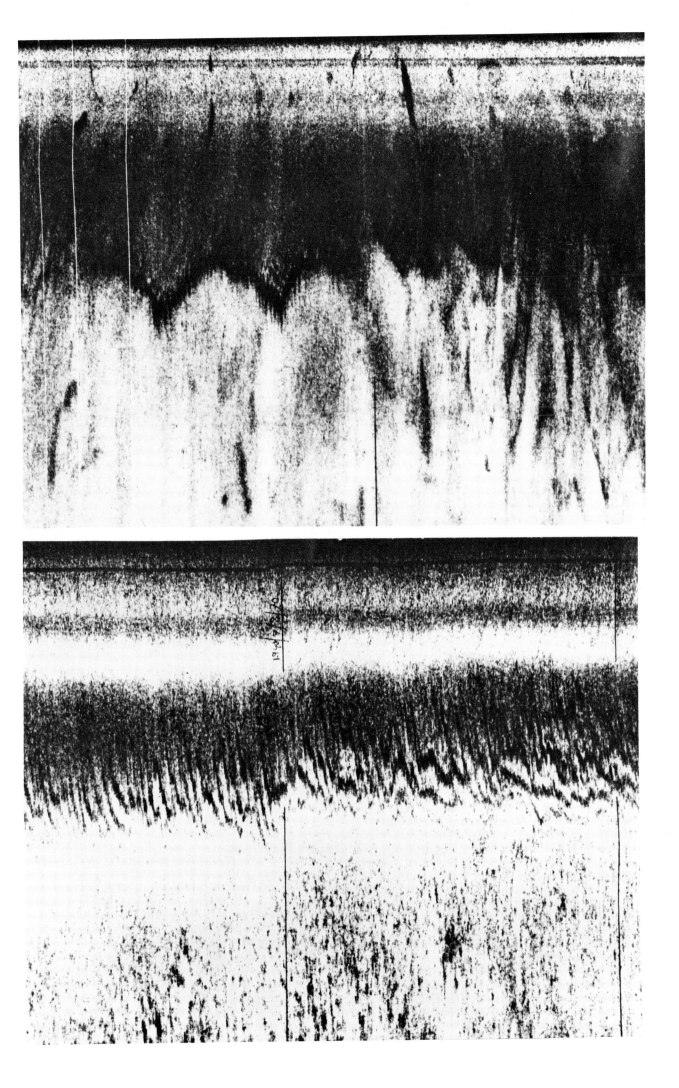

Ships

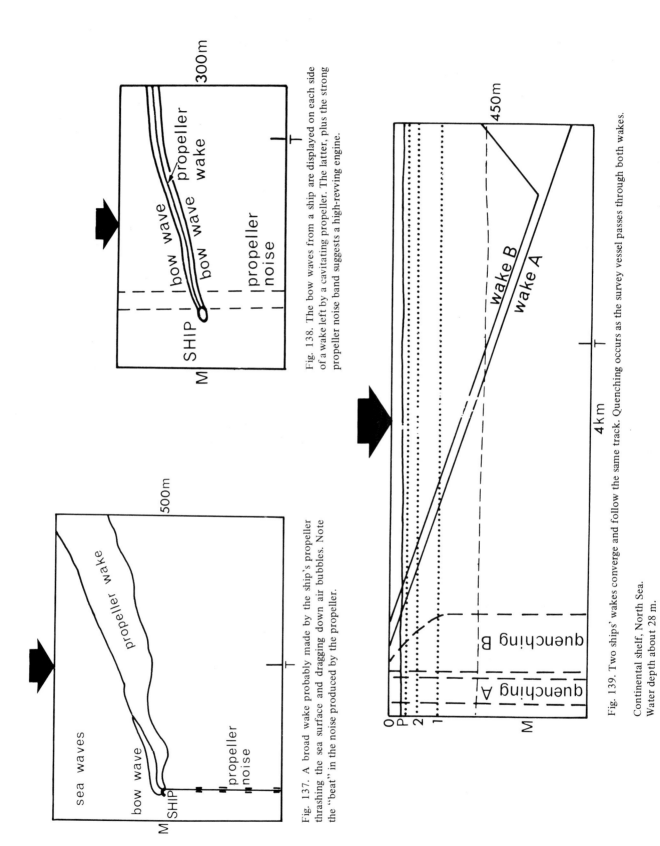

Fig. 137. A broad wake probably made by the ship's propeller thrashing the sea surface and dragging down air bubbles. Note the "beat" in the noise produced by the propeller.

Fig. 138. The bow waves from a ship are displayed on each side of a wake left by a cavitating propeller. The latter, plus the strong propeller noise band suggests a high-revving engine.

Fig. 139. Two ships' wakes converge and follow the same track. Quenching occurs as the survey vessel passes through both wakes.

Continental shelf, North Sea. Water depth about 28 m.

Ships

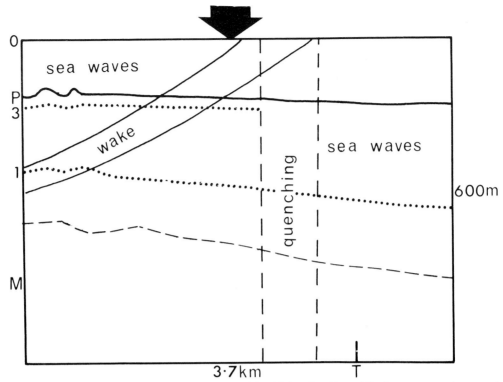

Fig. 140. Echoes were not obtained from the sea waves lying beyond the ship's wake. Also, while the survey vessel was passing through the wake the sea floor echoes were obscured by quenching.

Continental shelf, west of Portugal.
Water depth about 120 m.

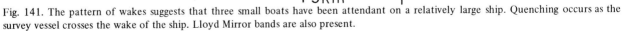

Fig. 141. The pattern of wakes suggests that three small boats have been attendant on a relatively large ship. Quenching occurs as the survey vessel crosses the wake of the ship. Lloyd Mirror bands are also present.

Continental shelf, Hong Kong.
Water depth about 10 m.
Towed Surveying Asdic. By courtesy of the Geophysics Group, Bath University.

160

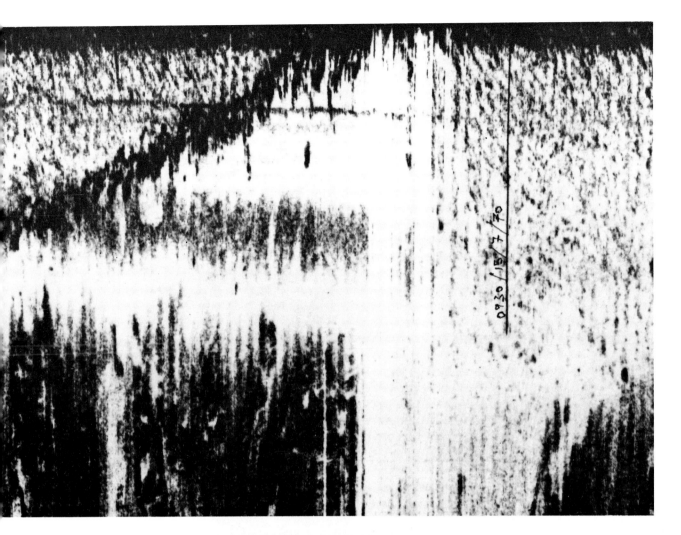

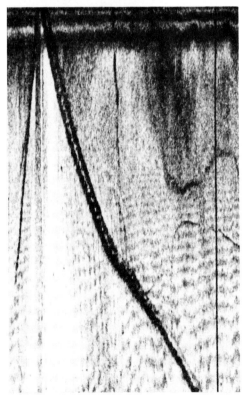

Wrecks

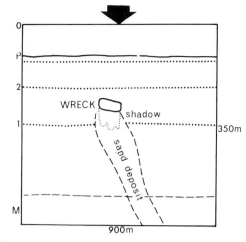

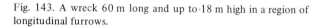

Fig. 142. A wreck 100 m long and up to 20 m high on a sandy floor with a sand accumulation on the lee side. The first side lobe is responsible for the detection of the wreck, shadow and most of the sand deposit.

Continental shelf, Celtic Sea.   51°32′N 05°13′W
Water depth about 65 m.

Fig. 143. A wreck 60 m long and up to·18 m high in a region of longitudinal furrows.

Continental shelf, English Channel.   50°22′N 02°03′W
Water depth about 45 m.

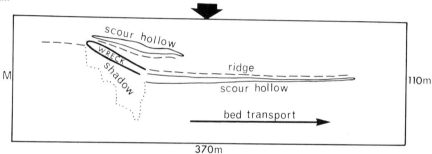

Fig. 144. A wreck 55 m long and up to 14 m high with ridges and scour hollows off each end. Their unequal development indicates the possible direction of net sediment transport.

Continental shelf, Dover Strait.
Water depth about 26 m.
Transit Sonar. By courtesy of Kelvin Hughes (Smiths Industries) Ltd.

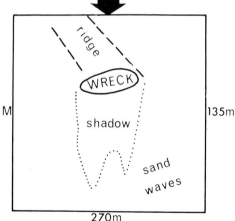

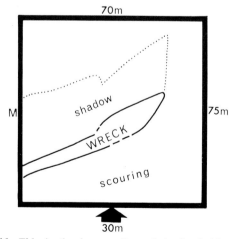

Fig. 145. A wreck 80 m long on a sandy floor. The sand waves and the deposit adjacent to the wreck indicate the probable direction of sediment transport.

Continental shelf, near Brest, France.
Water depth about 30 m.
Transit Sonar. By courtesy of Kelvin Hughes (Smiths Industries) Ltd.

Fig. 146. This is the bow section of the battleship H.M.S. "Formidable" that sank off Start Point in 1915. Tidal currents have produced scouring in front of the wreck (see sector scanning diagram between Fig. 148 and 149).

Continental shelf, English Channel.
Water depth about 60 m.
ARL Sector Scanning Sonar. By courtesy of Fisheries Laboratory, Lowestoft.

162

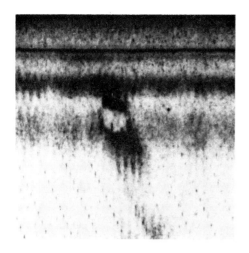

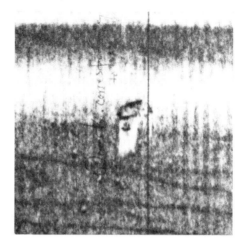

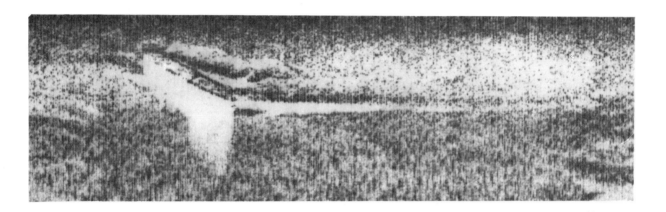

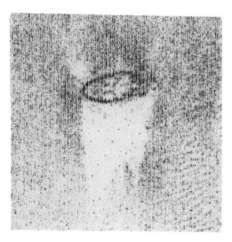

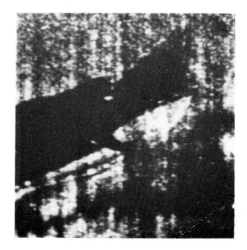

163

Bottom Marks

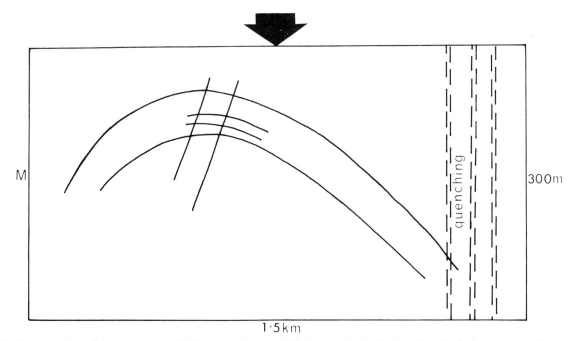

Fig. 147. Two sets of parallel marks on a muddy floor that have probably been made by the Otter Boards of a bottom trawl. In between the marks there is a broader mark possibly made by the Cod End.

Continental shelf, Firth of Clyde, Scotland.
Water depth about 40 m.

W.E. ⨯ 2.5

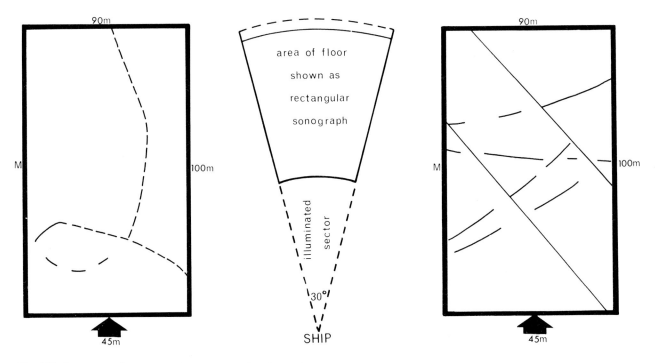

Fig. 148. Sector-scanning sonograph showing the looping track characteristic of an anchor suction dredger, which moves in short steps until favourable material is found.

Continental shelf, English Channel.   50°44′N 00°35′E
Water depth about 20 m.
ARL Sector Scanning Sonar. By courtesy of the Fisheries Laboratory, Lowestoft.

Fig. 149. Sector-scanning sonograph showing the criss-cross pattern of trenches due to continuous sampling by a trailing suction dredger. The trenches are about 5 m wide and the parallel pair are about 25 m apart.

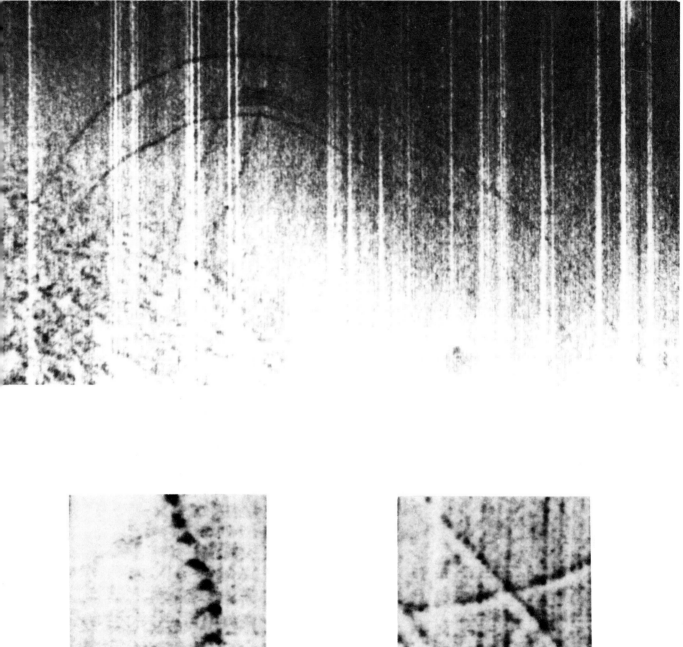

Cables

The extent to which a telephone or power cable is buried will depend on the composition of the sea floor and the water movements that exist in the vicinity.

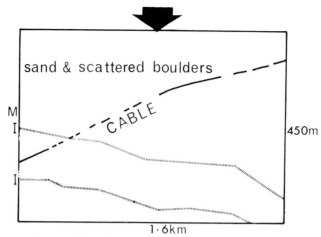

Fig. 150. A submarine cable. *I* indicates echo-sounder interference.

Continental shelf, Bristol Channel.
Water depth about 13 m.

51°18′N   03°44′W

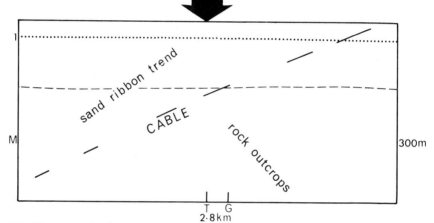

Fig. 151. A submarine cable. The saw-tooth effect is due to lack of stabilisation of the sound beams. At *G* there has been a change of amplifier gain.

Continental shelf, Bristol Channel.
Water depth about 30 m.

51°16′N   04°42′W

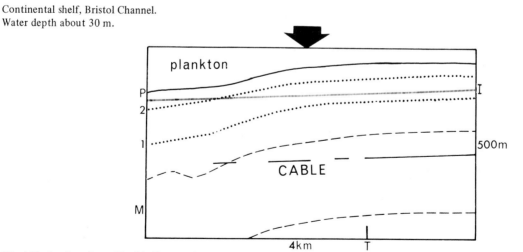

Fig. 152. A submarine cable. *I* indicates echo-sounder interference.

Continental shelf, Mediterranean.
Minimum water depth 260 m.

38°12′N   12°29′E

166

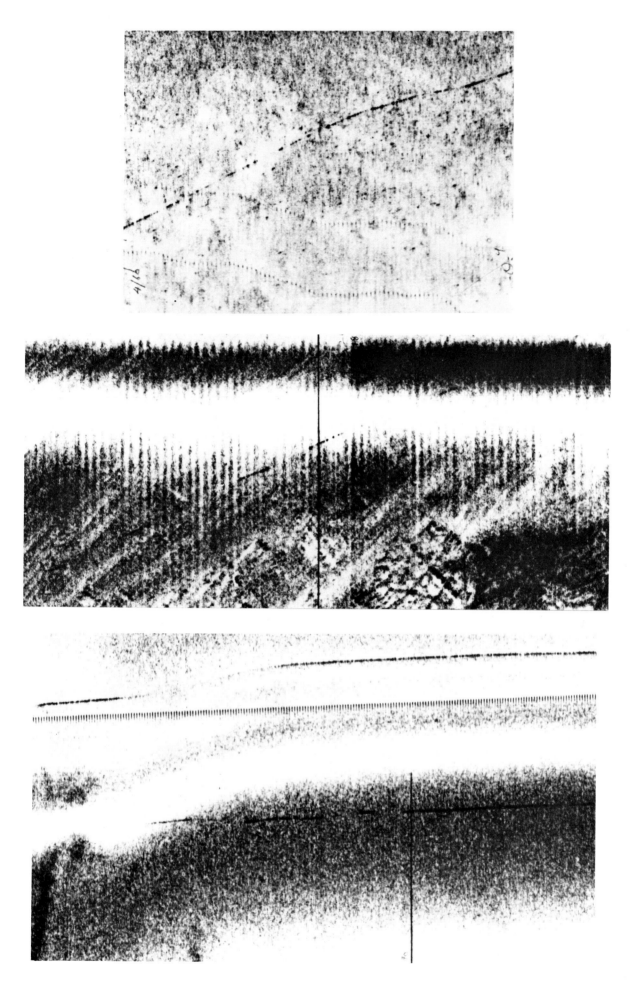

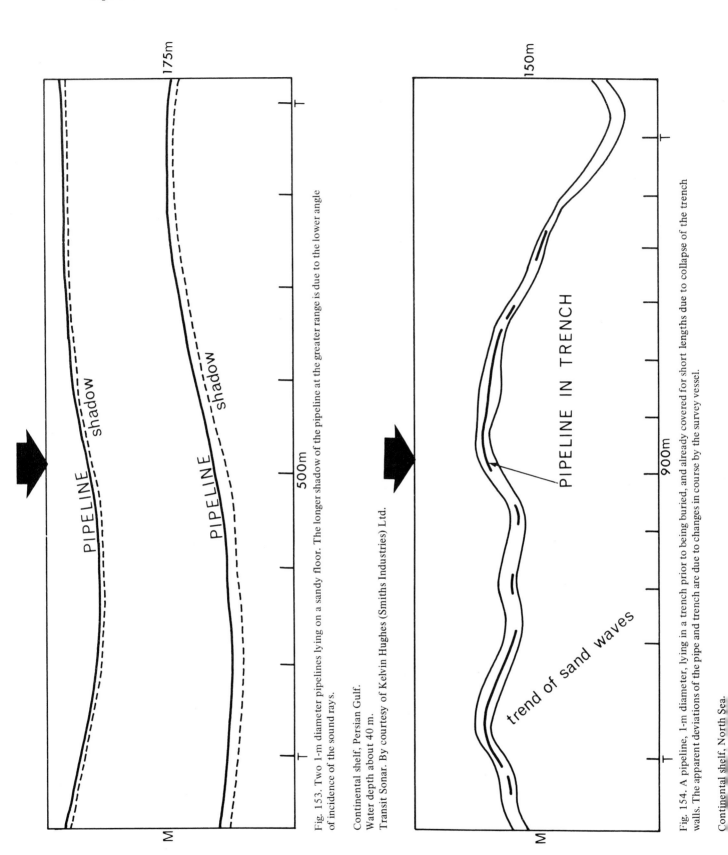

Fig. 153. Two 1-m diameter pipelines lying on a sandy floor. The longer shadow of the pipeline at the greater range is due to the lower angle of incidence of the sound rays.

Continental shelf, Persian Gulf.
Water depth about 40 m.
Transit Sonar. By courtesy of Kelvin Hughes (Smiths Industries) Ltd.

Fig. 154. A pipeline, 1-m diameter, lying in a trench prior to being buried, and already covered for short lengths due to collapse of the trench walls. The apparent deviations of the pipe and trench are due to changes in course by the survey vessel.

Continental shelf, North Sea.

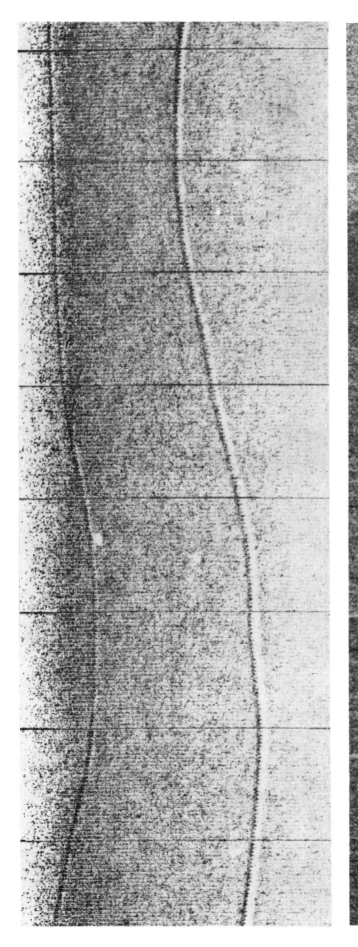

169

Pipelines

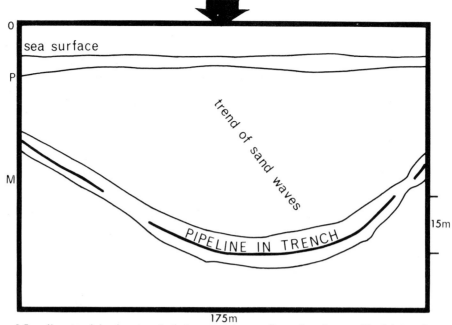

Fig. 155. A pipeline, 0.7 m diameter, lying in a trench that was cut in a sea floor of sand waves. The faint sand wave pattern appearing above the profile results from echoes, beyond the maximum range of a recorder sweep, which are displayed at the beginning of the next sweep. A sea surface profile is recorded because the transducer is towed at a depth of 8 m.

Continental shelf, North Sea.
Depth of water about 20 m.
Dual Side-scan Sonar. By courtesy of Decca Survey Ltd.

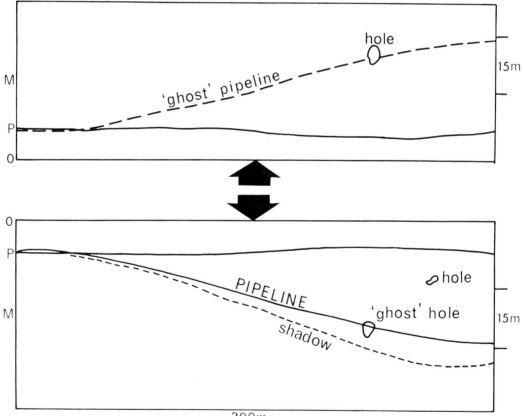

Fig. 156. A pair of sonographs obtained by looking to both sides simultaneously. The lower sonograph shows a pipeline on the sea floor. The shadow cast by the pipeline lengthens as the range increases because of the decreasing angle of incidence of the sound rays. The same frequency is used by both transducers and so the transducer facing the opposite direction has received the same pipeline echoes allowing a faint but spurious pipeline to be displayed on the upper sonograph. Similarly a hole in the sea floor on the upper sonograph is seen faintly on the lower one.

Continental shelf, Persian Gulf.
Depth of water about 7 m.
Dual Side-scan Sonar. By courtesy of Decca Survey Ltd.

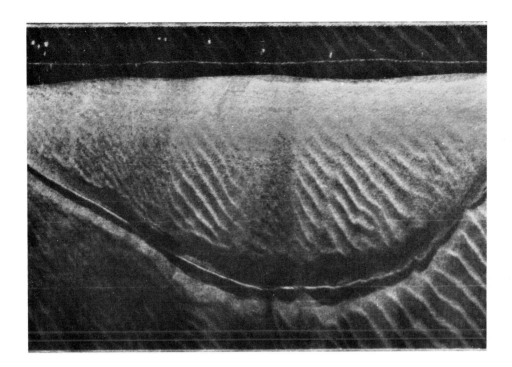

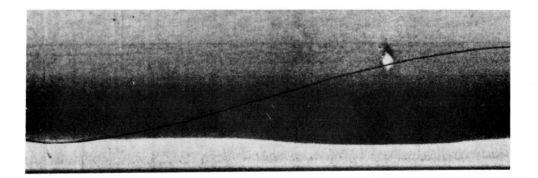

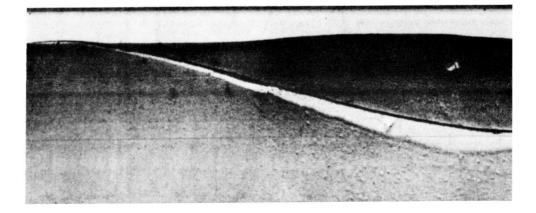

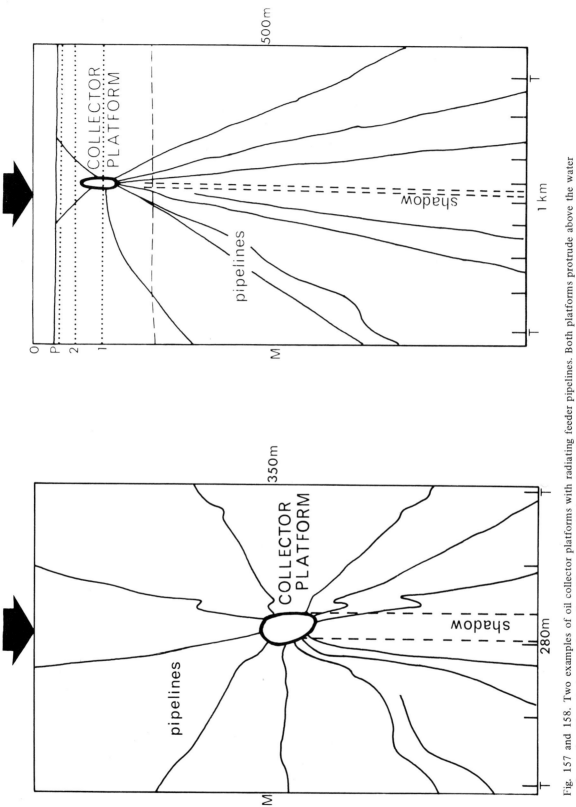

Fig. 157 and 158. Two examples of oil collector platforms with radiating feeder pipelines. Both platforms protrude above the water surface and cast long shadows on the sea floor. The apparent kinks in the pipelines are due to course deviations by the survey vessel.

Continental shelf, Persian Gulf.
Water depth about 25 m.
Towed Surveying Asdic. By courtesy of British Petroleum Co. Ltd.

172

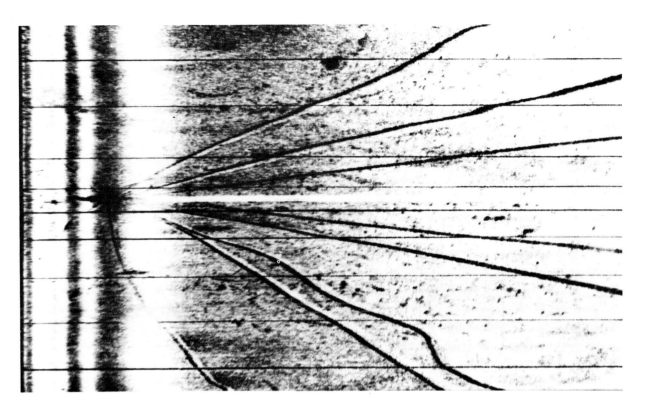

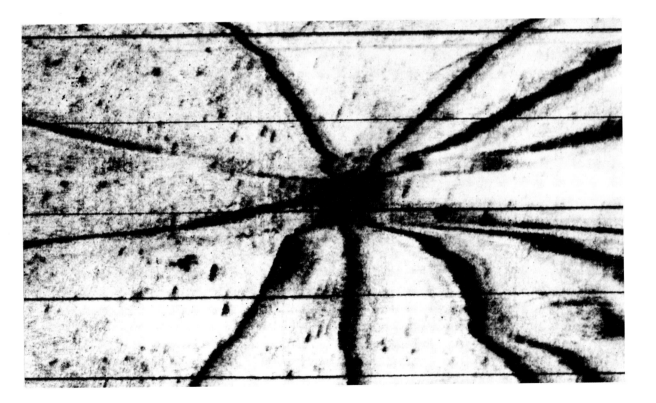

173

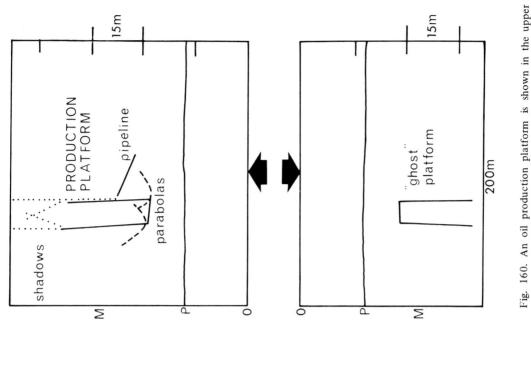

Fig. 160. An oil production platform is shown in the upper sonograph with shadows of the legs and bracing members. Very strong reflectors are responsible for the parabolic echo-patterns. The cross-coupling effect of the transducers has allowed a "ghost" platform to be recorded in the lower sonograph, and due to the reduced intensity, the component members of the rig are more easily distinguished.

Continental shelf, Persian Gulf.
Water depth about 18 m.

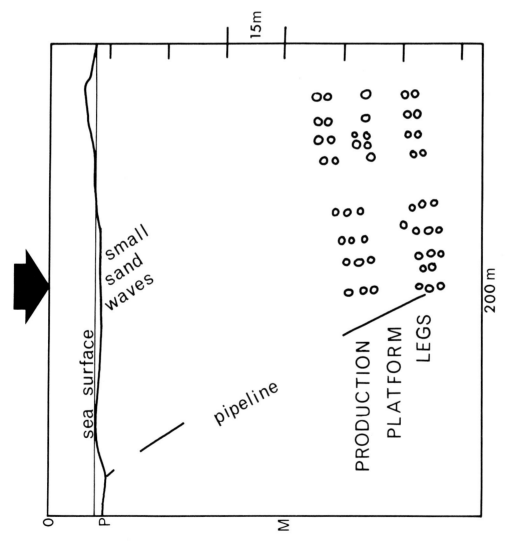

Fig. 159. Production platforms and a pipeline associated with the extraction of natural gas. The pipe-line is in a partially filled trench cut through sand waves. The transducer was towed at a depth of 15 m so that echoes from the sea surface and profile occur together.

Continental shelf, North Sea.
Water depth about 30 m.
Dual Side-scan Sonar. By courtesy of Decca Survey Ltd.

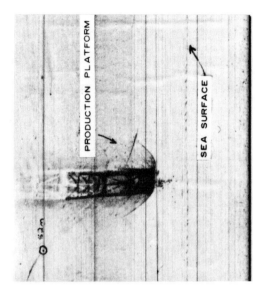

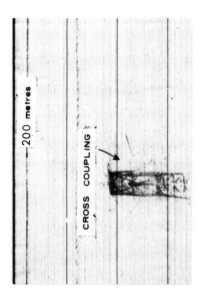

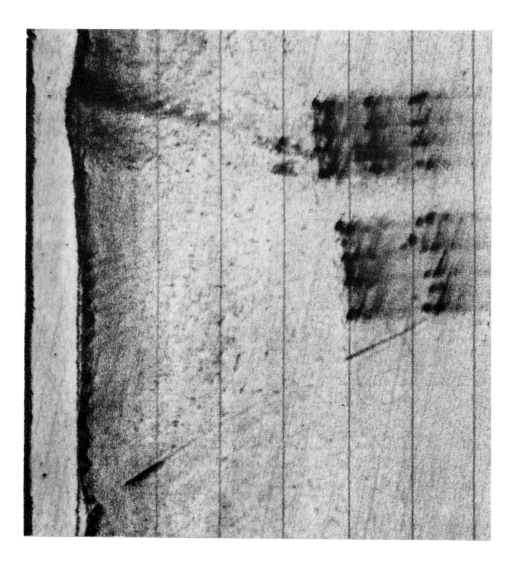

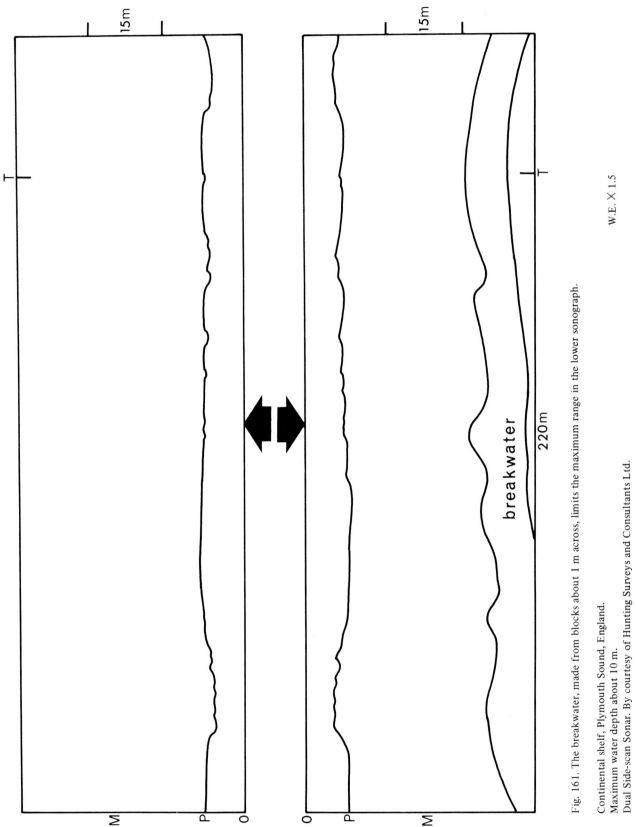

Fig. 161. The breakwater, made from blocks about 1 m across, limits the maximum range in the lower sonograph.

Continental shelf, Plymouth Sound, England.
Maximum water depth about 10 m.
Dual Side-scan Sonar. By courtesy of Hunting Surveys and Consultants Ltd.

Rock outcrops

Submarine cables

Acoustic shadow

STARBOARD CHANNEL

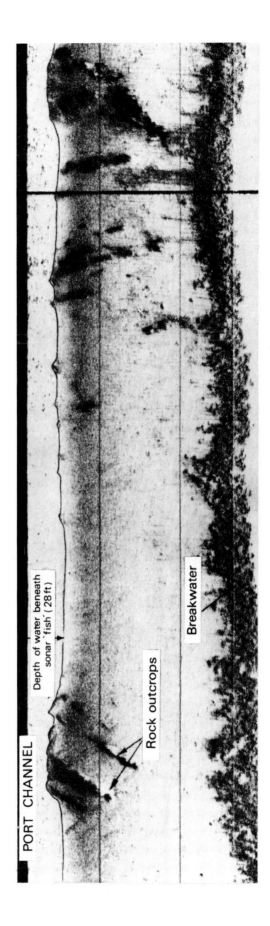

Depth of water beneath sonar 'fish' (28 ft)

Rock outcrops

Breakwater

PORT CHANNEL

Coastal Structures

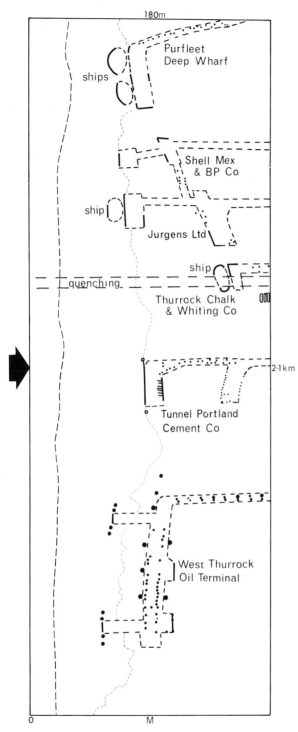

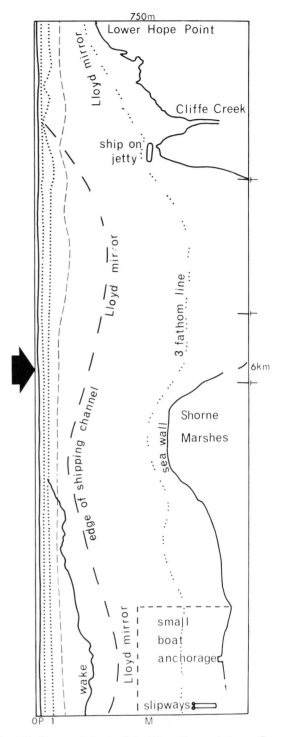

Fig. 162. The north bank of the River Thames between Purfleet and West Thurrock Marshes. The sonograph includes the change from deep to shallow water, the edge of the river being beyond the right hand side of the sonograph. The jetties are indicated on the sonograph by strong reflections from supporting piles (dots) and from flat surfaces (solid lines) facing the transducer. On the diagram these are surrounded by the outlines of the jetties (dashed) taken from an Admiralty chart. The breadth of the jetties and ships is increased by the exaggerated range scale of the sonograph. Objects on the river bed around the jetties add to the difficulty of determining their exact outlines.

Continental shelf, Thames Estuary, England.
Water depth about 9 m.   W.E. × 4
Transit Sonar. By courtesy of Kelvin Hughes (Smiths Industries) Ltd.

Fig. 163. The south bank of the River Thames between Gravesend and Lower Hope Point. A profusion of moorings, debris and river bed features are intermingled with small craft and wakes. Lloyd Mirror bands are apparent on the steep slopes of the dredged shipping channel. Information on the diagram not obtained from the sonograph has been taken from an Admiralty chart.

Continental shelf, Thames Estuary, England.
Water depth about 12 m.   W.E. × 2.5

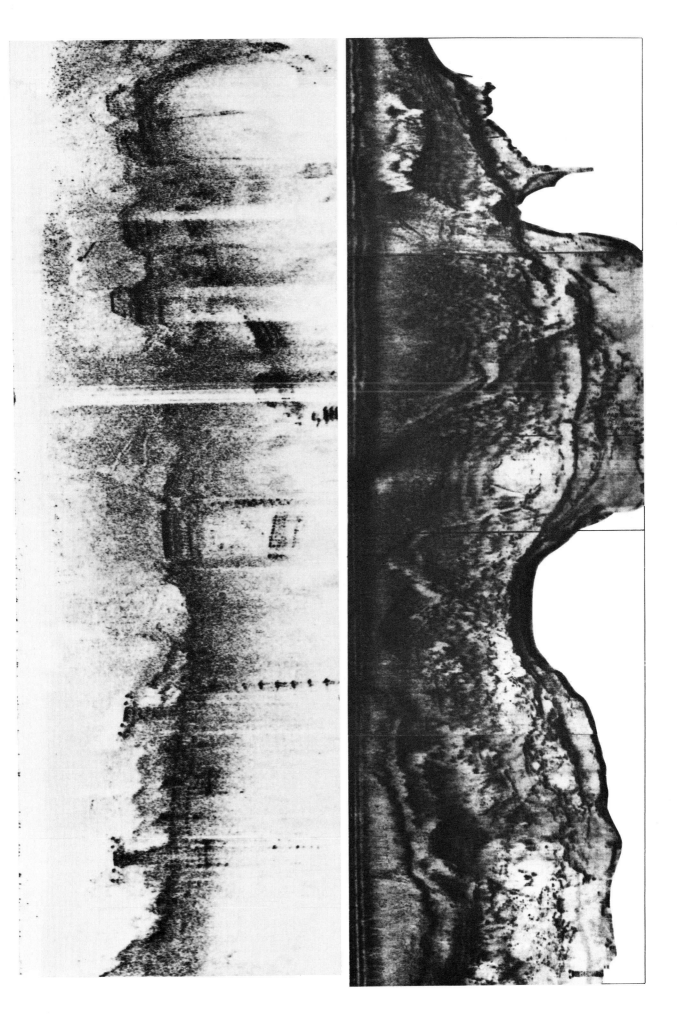

GLOSSARY

Bed: A layer of loose sediment or of sedimentary rock which can be extensive in area.

Bed form: A general term for such relief features as furrows and ridges, sand ribbons, sand waves and patches which indicate mobility or erosion of the sea floor materials by water movements. It also includes depositional bodies of sediment. Each is tied to a particular range of current speed and grain size, and for tidal seas each is probably related to the shape of the tidal ellipse.

Cavitation: The production of small transient voids on the lee side of a poorly streamlined body moving swiftly in water. Noise produced as the voids collapse can make marks on sonographs.

Continental shelf: The flattish sea floor surrounding the continents with characteristic depths down to about 200 m.

Continental slope: The long, relatively steep slope that lies between the outer edge of the continental shelf and the deep sea floor. Gradients are generally 3° to 6°, except where canyons, slumps and fault scarps are present.

Currents: The strengths quoted refer to average maximum value for spring tides near the sea surface, except in the case of the unidirectional Mediterranean undercurrent.

Dip direction: The direction of maximum slope of a bed of rock with respect to the horizontal.

Fault: A fracture plane along which there has been vertical and/or lateral displacement of one side with respect to the other.

Furrows and ridges: Longitudinal bed forms in gravel, sand or mud, some of which can be up to 9 km long, up to 14 m wide and may be as much as 1 m deep. They may be solitary but usually occur in groups, with separation of more than 25 m between furrows. The currents associated with them are strong — reaching speeds of about 3 knots (150 cm/sec) or more.

Gain: The increase (amplification) in the strength of the returning echoes as detected by the transducer that is required to display them on a recorder.

Gravel: Material coarser than sand, but finer than boulders. This general term includes the gravel, stones, pebbles, shingle and shell notations of navigational charts.

Gravel waves: Ridges of gravel, resembling sand waves, that generally occur in trains with crests transverse to currents that reach speeds of about 3 knots (150 cm/sec) and more. The examples shown in this book are generally about 10 m apart and up to about 1 m high.

Internal waves: Waves associated with a density gradient within the water and having considerable variation in height and wavelength. These complicate the effect of refraction and produce characteristic patterns on sonographs.

Joint: A fracture in rock without any noticeable vertical or horizontal displacement.

Lloyd Mirror effect: Light and dark bands extending along the sonograph, whose separation increases with distance from the ship and with increase in water depth. The pattern is only produced in shallow water during calm weather. It is caused by the interference between sound following a direct path from the sea floor to the transducer, with that following a longer route involving a reflection at the sea surface.

Longitudinal and transverse bed forms: Elongated parallel or transverse to the direction of the main current flow, respectively.

Main beam: The beam containing most of the sound energy emitted by a transducer (see schematic diagram on page 6). See also *Side lobes*.

Quenching: White lines across the sonograph, which result from air bubbles in the water blanketing sound transmission and reception from the transducer. This effect arises when a ship is pitching and rolling in bad weather.

Refraction: Bending of a sound ray occurs when it passes obliquely between water masses of different temperature and/or salinity. This distorting effect is very apparent on the sonographs where the upper part of the sea is warm, as in the tropics or in summertime elsewhere, and it can cause severe reduction of range and mask sea floor echoes.

Sand patches: Thinly spread bodies of sand no more than a few hundred metres across and commonly less than 2 m high, which may be depositional or subject to occasional movement. They usually extend transverse or parallel with tidal currents of less than about 1 knot (50 cm/sec), but their shape may be locally determined by the relief of the underlying floor.

Sand ribbons: Ribbons of sand can be up to 15 km long, 200 m wide and often are probably only a few cm thick. They are separated by coarser floor. Most of them are almost straight and parallel with currents reaching speeds of more than 2 knots (100 cm/sec).

Sand waves: Straight or sinuous ridges of sand commonly aligned transverse to currents of about 1–2 knots (50-100 cm/sec). They are analogous to some types of desert dunes. The crest separation can be between about 1 m and 1 km and heights reach up to about 20 m. Their cross section can be symmetrical but is generally asymmetrical, and is then a useful indicator of their migration direction. Sand waves of two or more sizes may occur together.

Seamount: An underwater mountain or hill generally of considerable size and often volcanic in origin.

Side-lobes: Secondary sound beams in which there is little energy (see schematic diagram, p. 6). Those occurring forward or aft of the main beam and those above it are of little consequence for side-scan sonar purposes and are not shown in the schematic diagram. In contrast, the ones below the main beam can be valuable for providing a profile of the sea floor as well as an oblique profile out to the side of the ship.

Slump: A downslope movement of material akin to a landslide, commonly occurring on continental or other slopes. They may be solitary or occur in groups, and may be small or involve appreciable areas and thickness of loose material.

Sonograph: An acoustic picture obtained under water by means of oblique 'illumination' with side-scan or sector-scanning sonars.

Strike: The line of outcrop in the horizontal plane of dipping layers of rock (at right angles to the direction of dip).

Submarine canyon: A large undersea valley, common on the continental slope, which resembles a deep gorge seen on land. Small channels in the sides of canyons are called gulleys.

Thermocline: The steepest part of the temperature gradient in a vertical section through the sea.

Tidal ellipse: A plot representing the changing speed and direction of the tidal current for a point in the sea throughout a tidal period, which may be about 12 or 24 h depending on locality.

Tilt angle: The angle between the mid-line of the main beam and the horizontal.

Transducer: A device which converts electrical energy into sound for the outgoing pulse and re-converts the incoming echoes back into electrical energy. The angular width and height of the main beam are largely governed by the width and height of the active face of the transducer, whereas the side-lobe pattern is governed by the shape and the distribution of sensitivity over the face of the transducer, and is therefore comparatively easy to control.

REFERENCES

References describing equipment are shown by a star*.

Ahrens, E., 1957. Use of horizontal sounding for wreck detection. *Int. Hydrogr. Rev.*, 34(2): 73–81.

* Anderson, V.C. and Lowenstein, C.D., 1968. Improvements in side-looking sonar for deep vehicles. In: F. Alt (Editor), *Marine Sciences Instrumentation*. Plenum, New York, N.Y., pp. 260–266.

* Anonymous, 1965. Sensor array for Thresher search. *Undersea Technol.*, 6(5): 31–33.

* Bass, G.F. and Rosencrantz, D.M., 1968. *A Diversified Program for the Study of Shallow Water Searching and Mapping Techniques*. Pennsylvania Univ., Philadelphia, Pa., N00014-67A-0216-0002, 130 pp.

Belderson, R.H., 1964. Holocene sedimentation in the western half of the Irish Sea. *Mar. Geol.*, 2: 147–163.

Belderson, R.H. and Kenyon, N.H., 1969. Direct illustration of one-way sand transport by tidal currents. *J. Sediment. Petrol.*, 39(3): 1249–1250.

Belderson, R.H. and Stride, A.H., 1966. Tidal current fashioning of a basal bed. *Mar. Geol.*, 4(4): 237–257.

Belderson, R.H. and Stride, A.H., 1969. The shape of submarine canyon heads revealed by Asdic. *Deep-Sea Res.*, 16(1): 103–104.

Belderson, R.H., Kenyon, N.H. and Stride, A.H., 1970. 10-km wide views of Mediterranean deep sea floor. *Deep-Sea Res.*, 17: 267–270.

Belderson, R.H., Kenyon, N.H. and Stride, A.H., 1971. Holocene sediments of the continental shelf west of the British Isles. In: F.M. Delany (Editor), *The Geology of the East Atlantic Continental Margin*. Institute of Geological Sciences, London, Rep. 70/14; pp. 161–170.

Brakl, J., Clay, C.S., Liang, W. and Fisher, G., 1969. An analysis of lateral echo-sounder records taken aboard the bathyscaphe "Archimede". *Hudson Lab., Columbia Univ., Tech. Rep.*, 170: 30 pp.

Breslau, L., Tittle, R., Krotser, D. and Fletcher, J., 1968. A riverbank echo-ranging system for riverine positioning. *U.S. Nav. Oceanogr. Off. Rep.*, IR 68–78: 20 pp.

Chesterman, W.D., 1968. Geophysical exploration of the ocean floor using acoustic methods. *Contemp. Phys.*, 9(5): 423–446.

Chesterman, W.D., 1970. *The Nature of the Ocean Floor. Inaugural Lecture*. Bath University of Technology, Bath, 26 pp.

* Chesterman, W.D., Clynick, P.R. and Stride, A.H., 1958. An acoustic aid to sea bed survey. *Acustica*, 8: 285–290.

* Chesterman, W.D., Quinton, J.M.P.S., Chan, Y and Matthews, H.R., 1967. Acoustic surveys of the sea floor near Hong Kong. *Int. Hydrogr. Rev.*, 44(1): 35–54.

* Cholet, J., Fontanel, A. and Grau, G., 1968. Etude du fond de la mer à l'aide d'un sonar latéral. *Inst. Fr. Pét. Rep.*, 15, 712: 18 pp.

* Clay, C.S. and Liang, W., 1964. Lateral echo-sounder model CL-1. *Hudson Lab. Columbia Univ., Tech. Rep.*, 114: 35 pp.

Clay, C.S., Ess, J. and Weismann, I., 1964. Lateral echo-sounding of the ocean bottom on the continental rise. *J. Geophys. Res.*, 69(18): 3823–3835.

* Clifford, P.J. and Henderson, R.F., 1968. Two developments in search and recovery. In: *A Critical Look at Marine Technology. Proc. Ann. Conf. Mar. Technol. Soc., 4th, Washington, D.C., 1968*, pp. 95–102.

Cook, J.C., 1971. Some practical applications and limitations of high definition depth scanning sonars. In: K.D. Troup (Editor), *Norspec 70, The North Sea Spectrum*. Thomas Reed, London, pp. 117–132.

* Daniels, C. and Henderson, R., 1969. An integrated acoustic underwater survey system. In: *Oceanol. Int. 69 Conf., Brighton, Great Britain, 1969, Tech. Sess., Day 4 – Underwater Observation and Communication*, 11 pp.

Decca Survey Ltd., 1969. Sea bed visualised with sonar. *Sci. J.*, 5(5): 32–33.

Dobson, M.R., 1969. The oblique asdic and its use in an investigation of a marine high-energy environment. *Sedimentology*, 13(1/2): 105–122.

Dobson, M.R., Evans, W.E. and James, K.H., 1971. The sediment on the floor of the Southern Irish Sea. *Mar. Geol.*, 11(1): 27–69.

Donovan, D.T. and Stride, A.H., 1961a. An acoustic survey of the sea floor south of Dorset and its geological interpretation. *Philos. Trans. R. Soc., B*, 244(712): 299–330.

Donovan, D.T. and Stride, A.H., 1961b. Erosion of a rock floor by tidal sand streams. *Geol. Mag.*, 98(5): 393–398.

Donovan, D.T., Savage, R.J.G., Stride, A.H. and Stubbs, A.R., 1961. Geology of the floor of the Bristol Channel. *Nature*, 189(4758): 51–52.

Dyer, K.R., 1970. Linear erosional furrows in Southampton Water. *Nature*, 225(5227): 56–58.

Eden, R.A., Carter, A.V.F. and McKeown, M.C., 1969. Submarine examination of Lower Carboniferous strata on inshore regions of the continental shelf of southeast Scotland. *Mar. Geol.*, 7(3): 235–251.

* Edgerton, H., 1968. Sound in the sea: pictures with sonar. *Tech. Eng. News*, 50(5): 7–13.

* E.G. and G. International, 1967a. *Side-Scan Sonar: Mark 1 Dual Channel System. Data Sheet OC 67–12*. 4 pp.

* E.G. and G. International, 1967b. A side-scan sonar plots the sea bed. *New Scientist*, 34(549): 648.

* Gaskell, T.F., 1965. Sideways facing asdic. *Le Pétrole et la Mer*, 1(103): 2 pp.

* Géomécanique, 1969. *The SOL–100*, 4 pp.

* Gilbert, R.L.G., Melanson, R.C., Eaton, R.M., Loring, D.H. and King, J., 1966. An experimental evaluation of the Kelvin Hughes horizontal echo sounder off the east Canadian coast. *Bedford Inst. Oceanogr., Intern. Note*, 66–4–I: 39 pp.

* Haines, R.G., 1963. Developments in ultrasonic instruments. *Int. Hydrogr. Rev.*, 40(1): 49–57.

Harden Jones, F.R. and McCartney, B.S., 1962. The use of electronic sector-scanning sonar for following the movements of fish shoals, sea trials of R.R.S. "Discovery II." *J. Cons., Cons. Perm. Intern. Exploration Mer*, (27(2): 141–149.

* Haslett, R.W.G., 1967. Some acoustic methods of underwater observation. In: *Br. Nat. Conf. Technol. Sea Sea Bed. U. K. At. Energy Auth. Rep.*, 20 pp.

Haslett, R.W.G., 1970. Acoustic echoes from targets under water. In: R.W.B. Stevens (Editor), *Underwater Acoustics*. Wiley, London, 129–197.

* Haslett, R.W.G. and Honnor, D., 1966a. Some recent developments in sideways-looking sonars. In: *Proc. I.E.R.E. Conf. Electron. Eng. Oceanogr., Southampton, 1966, Pap.*, 5: 11 pp.

* Haslett, R.W.G. and Honnor, D., 1966b. Simultaneous use of sideways-looking sonar, strata recorder and echo sounder. In:

Proc. I.E.R.E. Conf. Electron. Eng. Oceanogr., Southampton, 1966, Pap., 17: 8 pp.

Heaton, M.J.P., 1968. Profile, plan and section: three developments for sea-bed survey. Int. Hydrogr. Rev., 45(1): 73–80.

Heaton, M.J.P. and Haslett, W.G. 1971. Interpretation of Lloyd Mirror in side-scan sonar. Proc. Soc. Underwater Technol., 1(1): 24–38.

* Hopkins, J.C., 1970. Cathode ray tube display and correction of side-scan sonar signals. In: Proc. I.E.R.E. Conf. Electron. Eng. Ocean Technol, Swansea 1970, pp. 151–158.

Houbolt, J.J.H.C., 1968. Recent sediments in the southern bight of the North Sea. Geol. Mijnbouw, 47(4): 245–273.

* Institut Français du Pétrole, 1968. Lateral sonar – a recently developed technique for sea bottom reconnaissance. C.C.O.P. Tech. Bull., 1: 77–86.

* Kelvin Hughes, 1960. Fishermans Asdic Mark II. Publ. M360.

* Kelvin Hughes, 1962. Towed Surveying Asdic. Publ. KH 451.

* Kelvin Hughes, 1966. Transit SONAR (MS 43) Application. Publ. KH 1038a.

Kenyon, N.H., 1970a. Sand ribbons of European tidal seas. Mar. Geol., 9(1): 25–39.

Kenyon, N.H., 1970b. The origin of some transverse sand patches in the Celtic Sea. Geol. Mag., 107(3). 389–394.

Kenyon, N.H. and Stride, A.H., 1968. The crest length and sinuosity of some marine sand waves. J. Sediment. Petrol., 38(1): 255–259.

Kenyon, N.H. and Stride, A.H., 1970. The tide-swept continental shelf sediments between the Shetland Isles and France. Sedimentology, 14(2/3): 159–173.

King, L.H. and Maclean, B., 1970. Pockmarks on the Scotian Shelf. Geol. Soc. Am. Bull., 81(10): 3141–3148.

* Klein, M., 1967. Side scan sonar. Undersea Technol, 8(4): 24–26, 38.

Klein, M., 1971. Sonar search at Loch Ness. In: Preprints for 7th Annual Conference of Marine Technology Society. Marine Technology Society, Washington, D.C., pp. 423–430.

* Klein Associates, 1969. Side Scan Sonar Model MK–300. Klein, Salem, 18 pp.

* Kunze, W., 1957. General aspects of application of horizontal echo sounding method to shipping. Int. Hydrogr. Rev., 34(2): 63–72.

* Laing, J.T. and Nelkin, A., 1966. Seeing under the sea with sonar. Westinghouse Eng., 26(6): 162–168.

* Lee, O.S., 1961. Effect of an internal wave on sound in the ocean. J. Acoust. Soc. Am, 33(5): 677–681.

Loring, D.H. and Nota, D.J.G., 1966. Sea-floor conditions around the Magdalen Islands in the southern Gulf of St. Lawrence. J. Fish. Res. Board Can., 23(8): 1197–1207.

Loring, D.H., Nota, D.J.G., Chesterman, W.D. and Wong, H.K., 1970. Sedimentary environments on the Magdalen Shelf, southern Gulf of St. Lawrence. Mar. Geol., 8(5): 337–354.

Lowenstein, C.D., 1970. Side looking sonar navigation. J. Inst. Navig. 17(1): 56–66.

Luyendyk, B.P., 1970. Origin and history of abyssal hills in the northeast Pacific Ocean. Geol. Soc. Am. Bull., 81(8): 2237–2260.

McCartney, B.S., 1967. Underwater sound in oceanography. In: V.M. Albers (Editor). Underwater Acoustics. Plenum, New York, N.Y., 185–201.

McGehee, M.S., Luyendyk, B.P. and Boegeman, D.E., 1968. Location of an ancient Roman shipwreck by modern acoustic techniques. In: A Critical Look at Marine Technology. Proc. Ann. Conf. Mar. Technol. Soc. 4th, Washington, D.C., 1968, pp. 127–139.

* Mitson, R.B. and Cook; J.C., 1970. Shipboard installation and trials of an electronic sector scanning sonar. In: Proc. I.E.R.E. Conf. Electron. Eng. Ocean Technol., Swansea, 1970, pp. 187–210.

Mittleman, J.R. and Malloy, R.J., 1971. Stereo side-scan sonar imagery. In: Reprints for 7th Annual Conference of Marine Technology Society. Marine Technology Society, Washington, D.C., pp. 395–422.

Mudie, J.D., Normark, W.R. and Cray, E.J., 1970. Direct mapping of the sea floor using side-scanning sonar and transponder navigation. Geol. Soc. Am. Bull., 81: 1547–1554.

National Institute of Oceanography, 1969. GLORIA – successful first trials. Hydrospace, 2(3): 10–11.

Prentice, J.E., Beg, I.R., Colleypriest, C., Kirby, R., Sutcliffe, P.J.C., Dobson, M.R., D'Olier, B., Elvines, M.F., Kilenyi, T.I., Maddrell, R.J. and Phinn, T.R., 1968. Sediment transport in estuarine areas. Nature, 218(5148): 1207–1210.

Reeves, R.G., 1969. Structural geologic interpretations from radar imagery. Geol. Soc. Am. Bull., 80: 2159–2164.

Roberts, W.J.M., 1970. Large scale offshore surveying for the oil industry. Int. Hydrogr. Rev. 47(2): 41–64.

* Rusby, J.S.M., 1970. A long range side-scan sonar for use in the deep sea. Int. Hydrogr. Rev. 47(2): 25–39.

Rusby, J.S.M., Dobson, R., Edge, R.H., Pierce, F.E. and Somers, M.L., 1969. Records obtained from the trials of a long range side-scan sonar (GLORIA). Nature, 223(5212): 1255–1257.

Sanders, J.E., 1966. Geological calibration attempt of side-looking sonar, north shore of Minas Basin, Nova Scotia. Marit. Sediments, 2(1): 23–25.

* Sanders, F.H. and Stewart, R.W., 1954. Image interference in calm near-isothermal water. Can. J. Physics, 32: 599–619.

Sanders, J.E. and Clay, C.S., 1968. Investigating the ocean bottom with side-scanning sonar. In: J.E. Morgan and D.C. Parker (Editors), Proc. Symp. Remote Sensing Environment, 5th, University of Michigan, Ann Arbor, Mich., 1968, pp. 529–547.

Sanders, J.E., Emery, K.O. and Uchupi, E., 1969. Microtopography of five small areas of the continental shelf by side-scanning sonar. Geol. Soc. Am. Bull., 80: 561–572.

* Sapp, R.M., 1969. Chesapeake Bay bottom survey program. Westinghouse Oceanic Equipment ACE, Tech. Memo., 7: 25 pp.

Sargent, G.E.G., 1968a. Application of acoustics and ultrasonics to marine geology. Ultrasonics, 6(1): 23–28.

Sargent, G.E.G., 1968b. Profiling and sonar techniques for underwater pipeline surveying. Hydrospace, 1(5): 42–47.

Smith, A.J., Stride, A.H. and Whittard, W.F., 1965. The geology of the Western Approaches of the English Channel. IV. A recently discovered Variscan granite west-north-west of the Scilly Isles. Colston Pap., 17: 287–301.

* Somers, M.L., 1970. Signal processing in project GLORIA, a long range side scan sonar. In: Proc. I.E.R.E. Conf. Electron. Eng. Ocean Technol., 19: 109–120.

* Spiess, F.N., 1967. Instrumenting the environment – marine physics. In: Proc. Ann. Tech. Meet., 13th, Washington, D.C. Institute of Environmental Sciences, Washington, D.C. pp. 27–32.

Spiess, F.N. and Maxwell, A.E., 1964. Search for the Thresher. Science, 145(3630): 349–355.

* Spiess, F.N. and Mudie, J.D., 1970. Small scale topographic and magnetic features. In: A.E. Maxwell (Editor), The Sea, 4(1). Wiley, New York, N.Y., 205–250.

Spiess, F.N., Luyendyk, B.P., Larson, R.L. Normark W.R. and Mudie, J.D., 1969. Detailed geophysical studies on the northern Hawaiian Arch using a deeply towed instrument package. Mar. Geol., 7(6): 501–527.

Stephan, J.G., 1966. Mapping the ocean floor. *Battelle Tech. Rev., July-Aug.:* 7–11.

Stride, A.H., 1959. A linear pattern on the sea floor and its interpretation. *J. Mar. Biol. Assoc. U. K.,* 38(2): 313–318.

Stride, A.H., 1960. Recognition of folds and faults on rock surfaces beneath the sea. *Nature,* 185 (4716): 837.

Stride, A.H. 1961a. Geological interpretation of asdic records. *Int. Hydrogr. Rev.,* 38(1): 131–139.

Stride, A.H. 1961b. Mapping the sea floor with sound. *New Scientist,* 10: 304–306.

Stride, A.H., 1963. Current-swept sea floors near the southern half of Great Britain. *Q. J. Geol. Soc. Lond.,* 119: 175–199.

Stride, A.H., 1965a. Marine geology at the National Institute of Oceanography. *Times Sci. Rev.,* 58: 10–11.

Stride, A.H., 1965b. Under the North Sea. *Hunting Group Rev.,* 3: 20–23.

Stride, A.H., 1970. Mapping the ocean floor. *Sci. J.* 6(12): 56–59.

Stride, A.H., Belderson, R.H. and Kenyon, N.H., 1972. Longitudinal furrows and depositional sand bodies of the English Channel. *Bur. Rech. Geol. Minières,* 79: 233–244.

Stride, A.H., Curray, J.R., Moore, D.G. and Belderson, R.H., 1969. Marine geology of the Atlantic continental margin of Europe. *Philos. Trans. R. Soc., A,* 264 (1148): 31–75.

Stubbs, A.R., 1963. Identification of patterns on asdic records. *Int. Hydrogr. Rev.,* 40(2): 53–68.

Stubbs, A.R. and Lawrie, R.G.G., 1962. Asdic as an aid to spawning ground investigations. *J. Cons. Int. Exploration Mer,* 27(3): 248–260.

* Tucker, D.G., 1966. *Underwater Observation Using Sonar.* Fishing News (Books), London, 144 pp.

* Tucker, D.G., Welsby, V.G., Kay, L., Tucker, M.J., Stubbs, A.R. and Henderson, J.G., 1959. Underwater echo-ranging with electronic sector scanning: sea trials on R.R.S. "Discovery II." *J. Br. Inst. Radio Eng.,* 19(11): 681–696.

Tucker, M.J., 1961. Beam identification in multiple-beam echosounders. *Int. Hydrogr. Rev.,* 38(2): 25–32.

* Tucker, M.J., 1966. The use of sideways looking sonar for marine geology. *Geo-Mar. Technol.,* 2(9): 18–23.

* Tucker, M.J. and Stubbs, A.R., 1961. Narrow beam echo-ranger for fishery and geological investigations. *Br. J. Appl. Phys.,* 12: 103–110.

Tucker, M.J. and Stubbs, A.R., 1962. Underwater acoustics as a tool in oceanography. In: V.M. Albers (Editor), *Underwater Acoustics.* Plenum, New York, N.Y., 301–319.

Urick, R.J., 1967. *Principles of Underwater Sound for Engineers.* McGraw-Hill, New York, N.Y., 342 pp.

* Voglis, G.M., 1967. Underwater viewing with sonar. In: *Br. Natl. Conf. Technol. Sea Sea-Bed. U.K. At. Energy Auth. Rep.,* 7 pp.

Voglis, G.M. and Cook, J.C., 1966. Underwater applications of an advanced scanning equipment. *Ultrasonics,* 4: 1–9.

* Voglis, G.M. and Cook, J.C., 1970. A new source of acoustic noise observed in the North Sea. *Ultrasonics,* 8(2): 100–101.

* Walsh, G.M. and Moss, G.J., 1970. A new approach to preliminary site surveillance. *Navigation,* 17(2): 142–148.

* Westinghouse Electric Corporation. *Ocean Bottom Scanning Sonar.* Westinghouse Underseas Division, Baltimore, 6 pp.

Westinghouse Electric Corporation, 1964. Photographing the sea bed – by sound. *Engineering,* 197: 239–240.

Westinghouse Electric Corporation, 1970. Side-scan sonar swiftly surveys subsurface shellfish. *Westinghouse Perspective,* New York, N.Y., 4 pp.

Whitmarsh, R.B., 1971. Interpretation of long range sonar records obtained near the Azores. *Deep-Sea Res.,* 18: 433–440.

* Wong, H.K. and Chesterman, W.D., 1968. Bottom backscattering near grazing incidence in shallow water. *J. Acoust. Soc. Am.,* 44(6): 1713–1718.

Wong, H.K., Chesterman, W.D. and Bromhall, J.D., 1970. Comparative side-scan sonar and photographic survey of a coral bank. *Int. Hydrogr. Rev.,* 47(2): 11–23.

* Yules, J.A. and Edgerton, H.E., 1964. Bottom sonar search techniques. *Undersea Technol.,* 5(11): 29–32.